AUDITORY DEVELOPMENT IN INFANCY

ADVANCES IN THE STUDY OF COMMUNICATION AND AFFECT

Volume 1 • NONVERBAL COMMUNICATION
Edited by Lester Krames, Patricia Pliner, and Thomas Alloway

Volume 2 • NONVERBAL COMMUNICATION OF AGGRESSION
Edited by Patricia Pliner, Lester Krames, and Thomas Alloway

Volume 3 • ATTACHMENT BEHAVIOR
Edited by Thomas Alloway, Patricia Pliner, and Lester Krames

Volume 4 • AGGRESSION, DOMINANCE, AND INDIVIDUAL SPACING
Edited by Lester Krames, Patricia Pliner, and Thomas Alloway

Volume 5 • PERCEPTION OF EMOTION IN SELF AND OTHERS
Edited by Patricia Pliner, Kirk R. Blankstein, and Irwin M. Spigel

Volume 6 • ASSESSMENT AND MODIFICATION OF EMOTIONAL BEHAVIOR
Edited by Kirk R. Blankstein, Patricia Pliner, and Janet Polivy

Volume 7 • SELF-CONTROL AND SELF-MODIFICATION OF EMOTIONAL BEHAVIOR
Edited by Kirk R. Blankstein and Janet Polivy

Volume 8 • AGING AND COGNITIVE PROCESSES
Edited by F. I. M. Craik and Sandra Trehub

Volume 9 • INFANT MEMORY
Edited by Morris Moscovitch

Volume 10 • AUDITORY DEVELOPMENT IN INFANCY
Edited by Sandra E. Trehub and Bruce Schneider

ADVANCES IN THE STUDY OF
COMMUNICATION AND AFFECT

Volume 10

AUDITORY DEVELOPMENT IN INFANCY

Edited by

Sandra E. Trehub
and
Bruce Schneider
Center for Research in Human Development
Erindale College
Mississauga, Ontario, Canada

PLENUM PRESS • NEW YORK AND LONDON

Library of Congress Cataloging in Publication Data

Main entry under title:

Auditory development in infancy.

(Advances in the study of communication and affect; v. 10)
Based on the proceedings of a symposium held at Erindale College, University of Toronto in spring of 1981.
Includes bibliographies and index.
1. Auditory pathways—Congresses. 2. Auditory perception in children—Congresses. 3. Hearing disorders in children—Congresses. I. Trehub, Sandra, 1938- . II. Schneider, Bruce, 1941- . III. Series. [DNLM: 1. Auditory Pathways—growth & development—congresses. 2. Auditory Perception—in infancy & childhood—congresses. 3. Auditory Perception—physiology—congresses. 4. Hearing Disorders—in infancy & childhood—congresses. W1 AD8801 v.10/WV 272 A9126 1981]
RF291.5.C45A84 1985 618.92′0978 84-26422
ISBN 0-306-41757-X

A Division of Plenum Publishing Corporation
233 Spring Street, New York, N.Y. 10013

Printed in the United States of America

Contributors

W. KEITH BERG

Department of Psychology, University of Florida, Gainesville, Florida

P. A. BERNARD

Children's Hospital, 401 Smyth Road, Ottawa, Ontario

GORAN BREDBERG

Department of Audiology, Södersjukhuset, S–100, 64, Stockholm, Sweden

RACHEL KEEN CLIFTON

Department of Psychology, University of Massachusetts, Amherst, Massachusetts

HALLOWELL DAVIS

Central Institute for the Deaf, 818 South Euclid Street, St. Louis, Missouri

J. J. EGGERMONT

Department of Medical Physics and Biophysics, University of Nijmegen, Nijmegen, The Netherlands

REBECCA E. EILERS

Mailman Center of Child Development, Departments of Pediatrics and Psychology, University of Miami, Miami, Florida

DAVID M. GREEN

Department of Psychology, University of Florida, Gainesville, Florida

IVAN HUNTER-DUVAR

E. N. T. Department, The Hospital for Sick Children, 555 University Avenue, Toronto, Ontario

NEIL A. MACMILLAN

Department of Psychology, Brooklyn College, City University of New York, Brooklyn, New York

GEORGE T. MENCHER

Nova Scotia Hearing and Speech Clinic, Dalhousie University, Halifax, Nova Scotia

LENORE S. MENCHER

Co-ordinator Hearing Screening Program, Nova Scotia Hearing and Speech Clinic, Halifax, Nova Scotia

PHILIP A. MORSE

Department of Neuropsychology, New England Rehabilitation Hospital, Woburn, Massachusetts

Darwin W. Muir

Department of Psychology, Queen's University, Kingston, Ontario

D. Kimbrough Oller

Mailman Center for Child Development, Departments of Pediatrics and Psychology, University of Miami, Miami, Florida

Bruce A. Schneider

Centre for Research in Human Development, Erindale College, University of Toronto, Mississauga, Ontario

Sandra E. Trehub

Centre for Research in Human Development, Erindale College, University of Toronto, Mississauga, Ontario

Preface

The small but growing body of information about auditory processes in infancy is a tribute to the ingenuity and persistence of investigators in this realm. Undeterred by the frequent expressions of boredom, rage, and indifference in their subjects, these investigators nevertheless continue to seek answers to the intriguing but difficult questions about the course of auditory development.

In the spring of 1981, a group of leading scholars and researchers in audition gathered to discuss the topic, *Auditory Development in Infancy,* at the 11th annual psychology symposium at Erindale College, University of Toronto. They came from both sides of the Atlantic and from various disciplines, including audiology, neurology, physics, and psychology. They shared their views on theory and data, as well as their perspectives from the laboratory and clinic. One unexpected bonus was an unusually distinguished audience of researchers and clinicians who contributed to lively discussion within and beyond the formal sessions.

The principal goal of the symposium was to stimulate interdisciplinary communication on developmental issues in audition, particularly as these relate to the human infant. We would like to share the fruits of this endeavor with a wider audience by means of this volume, which contains edited versions of papers presented at the symposium. The papers are grouped in four parts, each concluding with a brief commentary. Part I concerns anatomical and physiological perspectives on auditory processes in infancy and the possible relations between underlying mechanisms and observed abilities. Contributors to Part II outline the development of some basic auditory abilities and describe behavioral approaches to the measurement of auditory sensitivity. In Part III, contributors focus on pathology, outlining demographic perspectives and the application of auditory-brainstem-response procedures to the diagnosis of auditory and nonauditory disorders. Lastly, in Part IV, attention is focused on complex auditory patterns, as well as models and mechanisms to account for the infants' processing of such patterns.

This volume and the symposium on which it is based would not have been possible without generous financial assistance from The Natural Sciences and Engineering Research Council of Canada, The Laidlaw Foundation, and Erindale College. We would also like to express gratitude to several individuals for their assistance in the organization and operation of the symposium: Dale Bull, Leigh Thorpe, Shannon Thompson, and Betty MacKenzie. Finally, we apologize to the contributors for unforeseen delays, primarily at the editorial level, in the publication of this volume.

SANDRA E. TREHUB

BRUCE A. SCHNEIDER

Contents

PART I. ANATOMY AND PHYSIOLOGY OF THE DEVELOPING EAR

PART IV. AUDITORY PATTERN PERCEPTION

PART I

ANATOMY AND PHYSIOLOGY OF THE DEVELOPING EAR

CHAPTER 1

The Anatomy of the Developing Ear

Goran Bredberg

Department of Audiology
Södersjukhuset
S–100 64, Stockholm
Sweden

The gross anatomy of the ear is shown in Figure 1. The ear is essentially a mechanical transducer of sound. Sound entering the external auditory canal vibrates the tympanic membrane or eardrum. The ossicles of the middle ear translate this vibratory motion of the eardrum into a pistonlike movement of the stapes at the oval window of the cochlea. The motion of the footplate of the stapes initiates a traveling wave along the basilar membrane of the cochlea. Distortion of the basilar membrane, in turn, exerts a shearing force on the hair cells, eventually leading to impulses along the eighth auditory nerve. Clearly then, auditory information processing will depend on, or at least be limited by, the mechanics of this system both at a gross-anatomical level (the resonance characteristics of the external auditory canal, the acoustic impedances of the various mechanical systems, etc.) and microanatomical level (the electrical and mechanical characteristics of the hair cells and supporting structures). Consequently, we might expect anatomical development to influence or limit how infants respond to sounds.

In the present chapter, we review the anatomy of the developing ear, giving special attention to the sensory mechanisms of the cochlear duct, indicating, whenever possible, how the process of development might influence how fetus and infant respond to sounds.

This work was supported by the Swedish Medical Research Council, Project No. 3542.

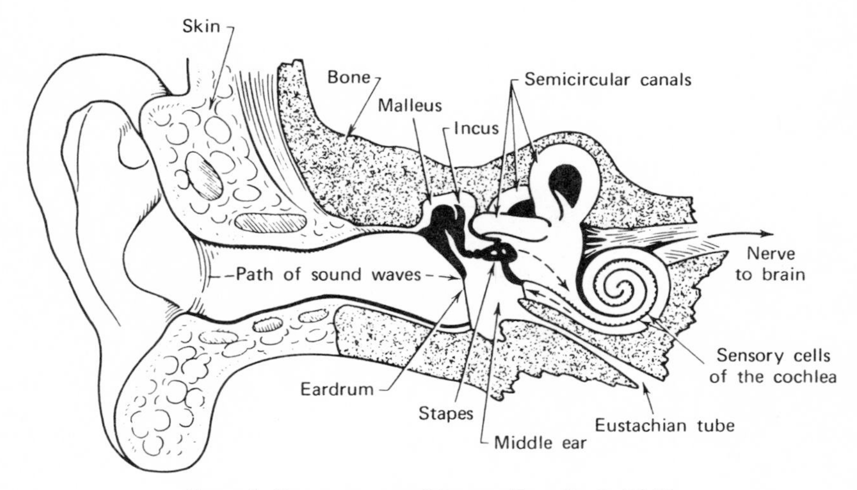

Figure 1. Gross anatomy of the ear. From Davis (1947).

The Gross Anatomy of the Ear

The auricle, external auditory canal, tympanic membrane, middle ear cavity, Eustachian tube, and ossicles of the middle ear all begin to form early in fetal development. Although many of these structures reach adult size and shape prior to birth, some, such as the auricle and the external auditory canal, do not reach their full size until adulthood. Changes in the size and shape of the auricle and auditory canal will affect the characteristics of the sound reaching the inner ear. Therefore, we might expect some differences between children and adults due to this rather slow growth rate in these external structures.

The structures of the middle ear reach their adult size and shape earlier than the external auditory canal. The middle ear cavity and the Eustachian tube arise from the first and probably the second pharyngeal pouches. The tube begins to be formed around the second month. At 10 weeks, the tube is largely formed, and the middle ear consists of a narrow cleft that slowly widens and contacts the external auditory meatus. The antrum is formed around this age, and the mastoid air cells begin to form late in fetal life; this development continues during infancy and childhood.

The growth of the ossicles, on the other hand, is so rapid that they have attained their adult size and general shape by around 15 weeks. Bone formation is initiated at 16 weeks and is finished at approximately 25 weeks. Both the malleus and the incus do not change their general form during this process. The stapes, however, has a very different shape in the early fetal stage, being an irregular cartilaginous ring that becomes ossified and much thinner, so that in the late fetal and in the newborn periods the stapes has almost attained its mature structure.

The structures of the inner ear or membranous labyrinth arise as an invagination of the ectoderm. The earliest manifestations are thickenings of the ectoderm with one appearing on either side of the brain. These thickenings, or auditory placodes, become invaginated to form auditory sacs, which are later separated from the surface to form the auditory vesicles, or otocysts, in the 4- to 6-mm human embryo. Three infoldings of the wall of an otocyst later subdivide it into the principal structures of the inner ear: (1) the endolymphatic duct and sac, (2) the utricle and semicircular canals, and (3) the saccule with its outgrowth, the ductus reuniens and the cochlear duct. Further details concerning the development of these structures may be obtained from the monographs by Bast and Anson (1949) and Bredberg (1968). In the remainder of this chapter, we will concentrate on the development of the sensory epithelium of the cochlear duct, since it is the development of this portion of the auditory system that is likely to have the most profound effect on the infant's ability to process auditory signals.

The Cochlear Duct

A cross section of the adult cochlear duct is shown in Figure 2. The cochlear duct is a helical tube that develops as an outgrowth from the anteromedial end of

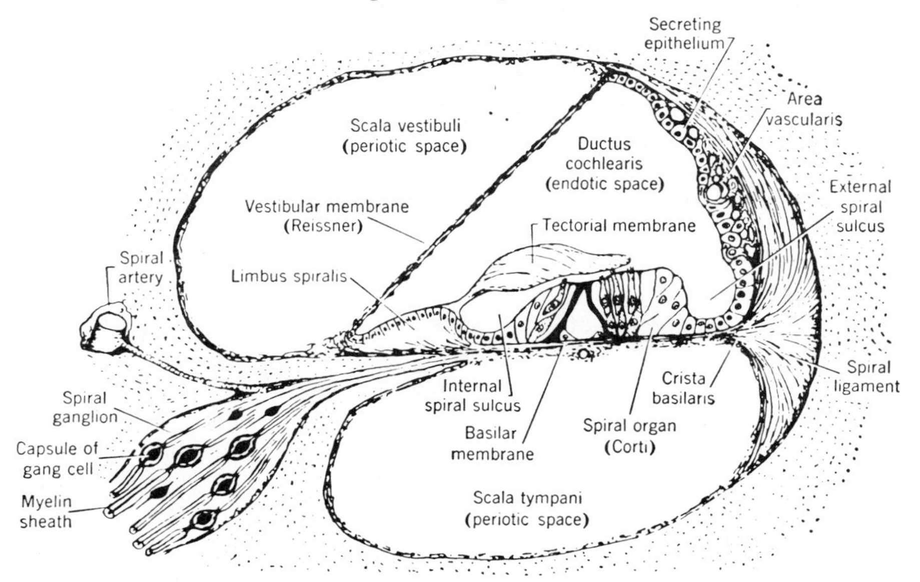

Figure 2. Cross section of the cochlea. From Rasmussen (1943).

the saccular portion of the otocyst. In young fetuses up to the age of 8 weeks, it is the only part of the cochlea present. The scala vestibuli, the scala tympani, the modiolus, and the bony capsule develop later from the surrounding mesenchyme.

The cochlear duct in the 7-week fetus has developed into a curved tube that describes one turn. At the 30-mm stage (8½ weeks), it has increased to one and three-quarter turns. At the 50-mm stage (10 to 11 weeks), the definitive two and three quarter turns are present. Beyond this stage, the spiral grows in size but not in the number of turns, reaching its maximum size at about midterm (Bast & Anson 1949).

Anatomical Development of the Organ of Corti

The differentiation of the sensory epithelium into the organ of Corti does not proceed uniformly throughout the coils of the cochlea. Most authors have stated that the development begins at the basal end and proceeds gradually toward the apex (Alexander, 1926; Bast & Anson, 1949; Held, 1926; Kolmer, 1927; Retzius, 1884; Van der Stricht, 1919, 1920; Wada, 1923; Weibel, 1957). Lorente de No (1933), however, described a somewhat different pattern in the mouse in which differentiation began at a point a short distance from the base and progressed in both directions from there. This suggests that the middle portion is more likely to mature and begin functioning earlier than either the basal or apical extremes. Similarly, Larsell, McCrady, and Larsell (1944) found that, in the opossum, morphological differentiation in the upper basal and lower middle coils preceded differentiation in the basal and apical extremities. They also found that the young opossum in the pouch reacted earliest to sound stimuli in the relative narrow frequency band of the middle range. Later, the opossum began to react to lower and higher frequencies, again suggesting that the middle portion matures first. In the rabbit, Änggård (1965) electrophysiologically showed that the earliest sign of cochlear function occurred in response to stimulation with medium to high frequencies (2–5 kHz). Similar observations have been made in the rat (Crowley & Hepp-Reymond, 1966) and in the mouse (Mikaelian & Ruben, 1965).

Ruben (1967) examined cochlear development by studying terminal mitoses of the cells in the membranous labyrinth. Terminal mitoses are the last divisions that a cell undergoes and thus serve as an index for the time of establishment of a permanent cell population. Ruben found that the hair cells (as well as pillar cells and Deiters' cells) in the cochlea were distributed in such a way that the oldest cells, the cells that undergo terminal mitoses first, were at the apex; and the youngest cells, the cells that undergo terminal mitosis last, were at the base. Ruben suggested that the growth area in the organ of Corti might be at the basal end and that growth by cell division pushes those cells whose cell division is completed toward the apex. This would appear to imply that, although the apical cells reach terminal mitosis first, they take longer to reach their mature form than the relatively younger hair cells located in the basal turn.

In the beginning of the third month, the cochlear duct has, in cross section, an oval shape that later becomes triangular when Reissner's membrane begins to separate from the surrounding mesenchyme (Figure 3).

In the 3-month (9 cm) human fetus, the cochlea has formed all its turns but has not reached its full size. (The length of the cochlear duct is 20 mm.) The scala vestibuli and tympani have begun to differentiate from the surrounding

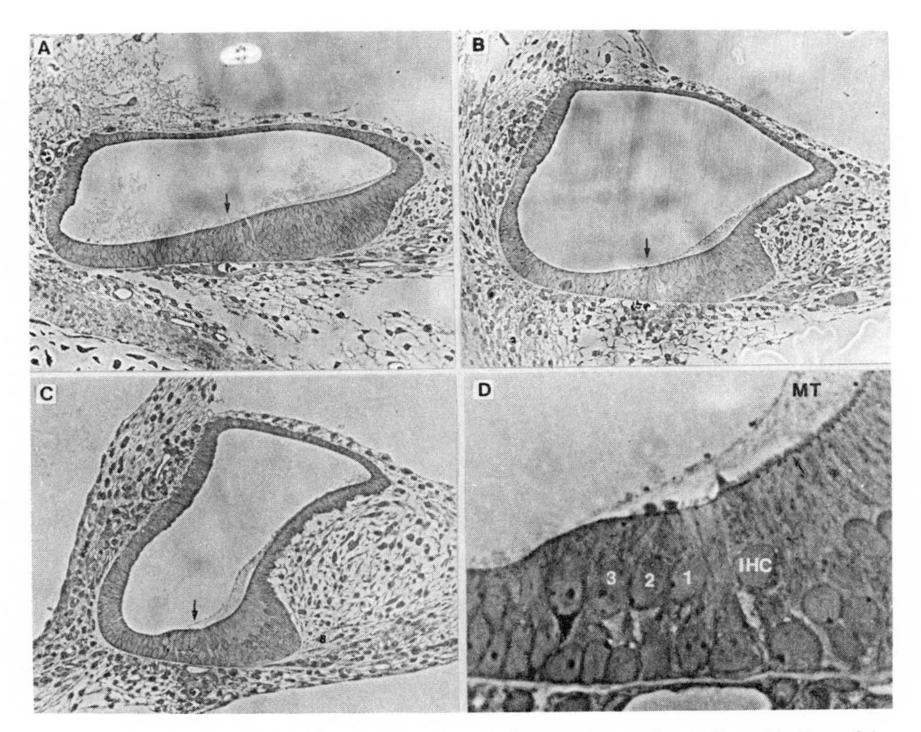

Figure 3. Cross sections through the cochlear duct of a 3-month human fetus. The epithelium of the cochlear duct stands out clearly from the mesenchymal tissue. (A) One and a half coils (15 mm) from the basal end, the scala vestibuli (above) is more clearly developed than the scala tympani. The hair cell area is indicated with an arrow (× 100). (B) Three quarters of a coil (9 mm) from the basal end the hair cells are discernible (arrow) (× 100). (C) 3 to 4 mm from the basal end the hair cells (arrow) can be clearly differentiated (× 100). (D) Same areas as in (C), at a higher magnification, showing one inner hair cell (IHC) and three outer hair cells (1, 2, 3). Hairs can be seen on both inner and outer hair cells. Note the microvilli (arrow) on the supporting cells under the tectorial membrane (MT) (× 520).

mesenchyme, the development of the former being more advanced than that of the latter. The future organ of Corti consists of a thickening of the epithelium, which has formed two spirally running ridges at the base of the cochlea. As seen from above in a surface specimen, most of the epithelium show a characteristic, uniform pattern of polygonal cell surfaces; however, in a narrow region of the epithelium, at the junction of the inner and outer ridges, cells are organized in spirally running rows, indicating the region of inner and outer hair cells (Figure 4). Light and dark cells can be distinguished. The dark cells differentiate to become hair cells, the light cells to become supporting cells. Kolmer (1927) suggested that by this stage no further mitotic divisions occur, and that, therefore, subsequent growth must occur by cell growth and further differentiation. The inner hair cells show earlier differentiation than the outer.

About 3 to 4 mm from the basal end of the cochlea, the outer hair cells are disposed in three regular rows and have a round, free upper surface, whereas the inner hair cells lie in a single row and present a larger, free upper surface. The inner and outer hair cells are separated by the inner and outer pillar cells; at the free surface of the epithelium, however, the inner pillar heads alone are represented. On the free surfaces of sensory cells, as well as on the supporting cells, a kinocilium is present, already visible in the phase-contrast microscope as a distinct, dark dot. Stereocilia are present on both inner and outer hair cells (Bredberg, 1968; Bredberg, Ades, & Engström, 1972; Igarashi & Ishii, 1980; Wersäll & Flock, 1967). From this region toward the basal end of the organ of Corti, both inner and outer hair cells become smaller (Figure 4). The inner hair cells, arranged in a regular row, are distinct and clearly visible to the end of the organ of Corti, whereas the outer hair cells are irregularly arranged in two or three rows, and, in the last 10 mm, are indistinguishable from the supporting cells.

Apicalward of the differentiated region, a similar gradual change of pattern is seen. At a point one and a half coils (15 mm) from the base, neither outer nor inner hair cells can be differentiated at the free surface of the epithelium.

Intraepithelial fluid spaces were not observed at any level of the cochlea at this fetal age.

The mesenchymal plate, beneath the cochlear duct, is richly vascularized. The most prominent vessel runs spirally below the junction of the two epithelial ridges (Figure 5). In the basal coil it measures 20 to 40μ in diameter. It is continuous from base to apex, although its diameter diminishes in the apical coil. Vessels from the modiolus run outward to the spiral vessel, the latter sending branches toward the periphery. Some vessels run from the modiolus directly to the periphery without a connection to the spiral vessel. Vessels crossing the future basilar membrane occur more frequently in the apical coil.

At 4 months (14–17 cm), the scala tympani and the scala vestibuli are well developed and the organ of Corti has reached its mature length. When compared

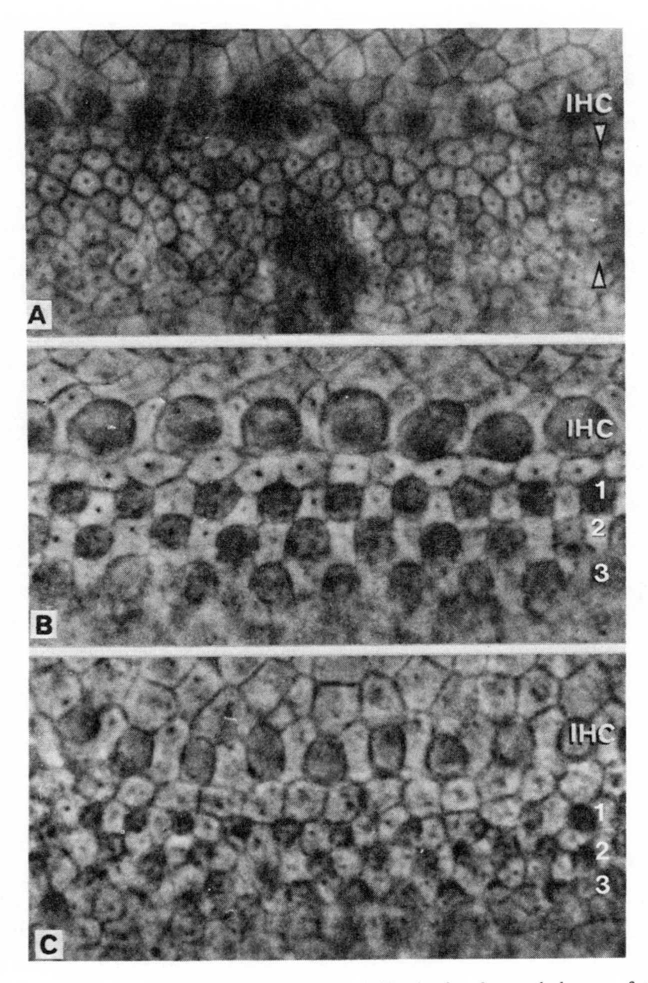

Figure 4. Three surface preparations of the organ of Corti of a 3-month human fetus, showing different stages of development at different levels of the cochlea. (A) One coil (11 mm) from the basal end of the cochlea the inner hair cells (IHC) can be distinguished by their dark appearance. However, there is as yet no sign of differentiation of the sensory and supporting cells in the area of the future outer hair cells (between arrows) (× 1280). (B) In this region, 3 to 4 mm from the basal end, both inner (IHC) and outer (1, 2, 3) hair cells are differentiated (× 1280). (C) 1 mm from the basal end, the surface area of the sensory cells is smaller. The inner hair cells (IHC) are well seen, but some of the outer hair cells are difficult to distinguish from the supporting cells (× 1280).

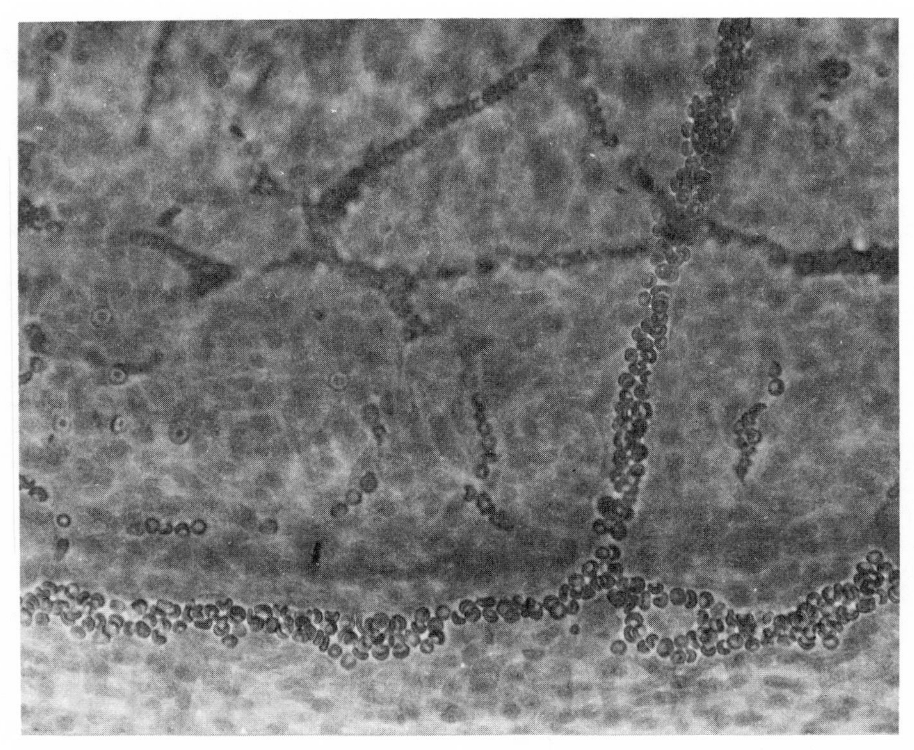

Figure 5. The future osseus spiral lamina and the basilar membrane from the middle coil of a 3-month human fetus. The blood cells in the vessels are visualized. Note the thick vessel running from the modiolus (above) to the area of the inner and outer ridges (organ of Corti) below (× 200).

to the 3-month fetus, the surface pattern has changed considerably (Figure 6). The inner hair cells can be distinguished throughout all the coils of the cochlea. Occasional supernumerary inner hair cells are distributed throughout the coils.

Outer hair cells can be recognized from the basal end to within a few millimeters of the apex. In the most basal 2 mm, they are distributed irregularly in two or three rows. Close to apex, the outer hair cells cannot be differentiated from the supporting cells.

Although, in general, the pattern of outer sensory cells is very regular in the basal and middle coils, occasional irregularities are present. In most cases, the irregularity is in the form of single supernumerary cells. A few cells are occasionally found forming a fourth row. At this stage, the sterocilia on the sensory cells display the characteristic W-shaped pattern. Fluid spaces in the organ of Corti are still not seen. The blood vessels show no striking difference from the 3-month stage.

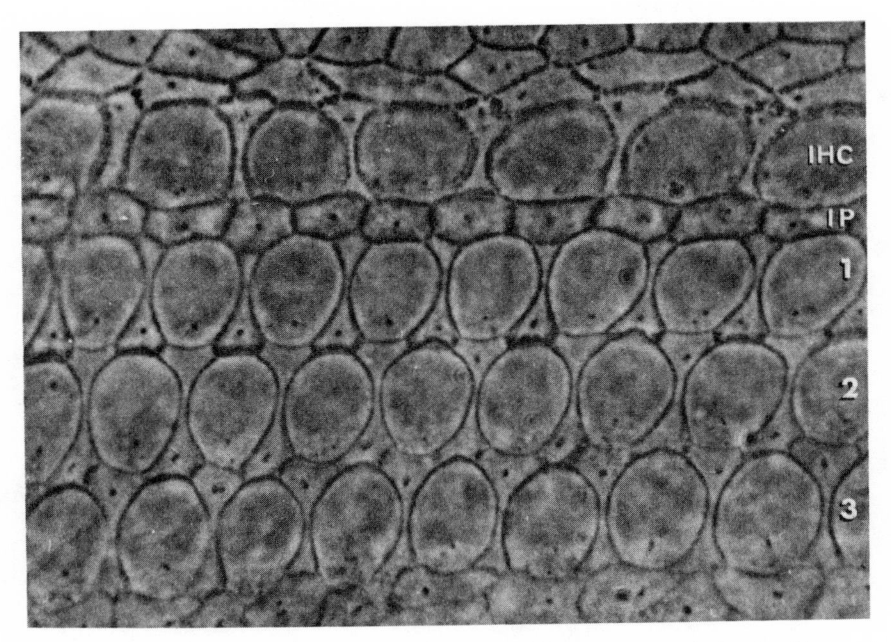

Figure 6. Surface preparation of the organ of Corti. Half a coil (11 mm) from the basal end from a 4-month human fetus. Both inner and outer (1, 2, 3) hair cells are well developed. At this age the cell pattern has an almost geometric regularity. A kinocilium is evident as a dark dot on both sensory and supporting cells (× 1300).

At 5 months (22–23 cm), the upper surfaces of the outer hair cells of the basal and middle coils have acquired the shape generally characteristic of mature, outer hair cells; however, in the apical coil, their free surfaces remain smaller and more rounded. The free surfaces of the inner hair cells show the mature shape throughout all coils. In general, the outer hair cells are disposed in three regular rows, although cells forming a fourth row are seen more frequently than at earlier stages, especially in the upper coils. At this age, intraepithelial fluid spaces are evident in phase-contrast microscopy (Figure 7).

The acoustic papilla grows rapidly in size as the intraepithelial fluid spaces are formed and widened. The formation of the fluid spaces has been studied in animals especially by Van der Stricht (1919) and Weibel (1957). The initially narrow intercellular clefts expand to become relatively wide intraepithelial spaces, while the supporting cells change in shape to become more slender. Van der Stricht (1919) considered the fluid spaces to be formed by a reduction of the volume of the supporting cells through cytolysis and liquefaction. The subsequent widening of Nuel's space cannot be explained by diminishing cell bodies,

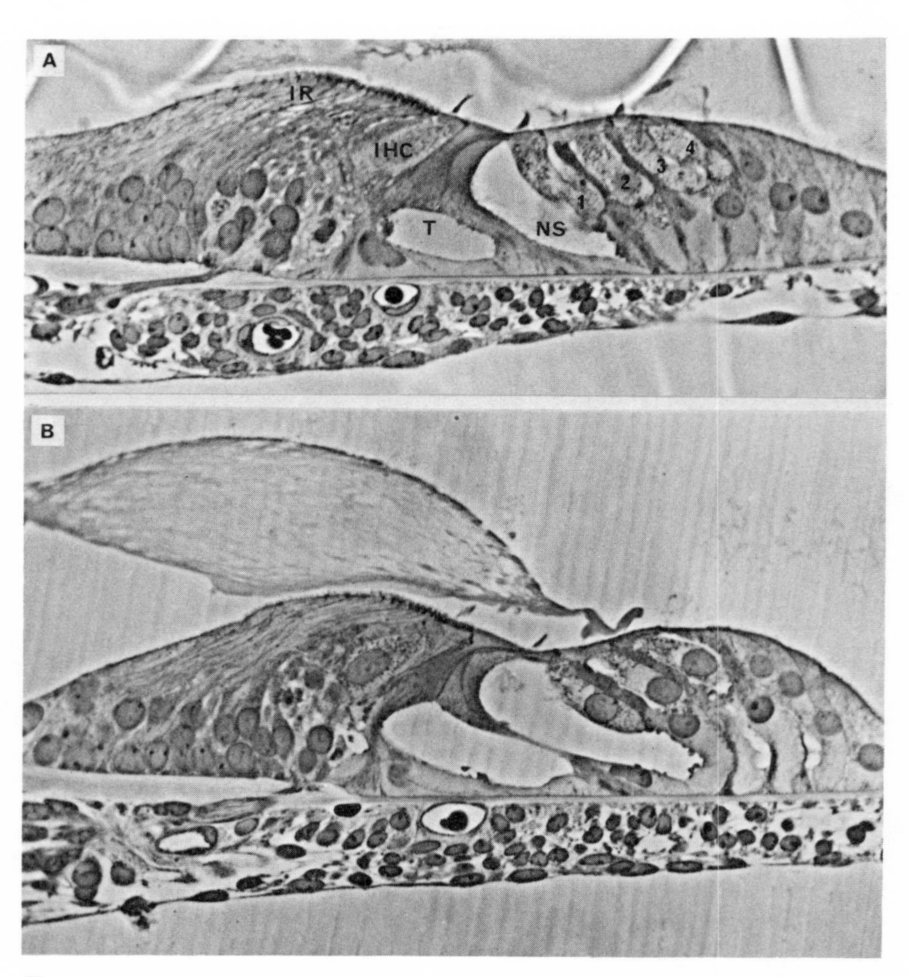

Figure 7. Cross sections through the organ of Corti of a 21-week human fetus showing different stages of development of the fluid spaces at different levels of the cochlea. (A) One coil (21 mm) from the basal end, the tunnel of Corti (T) is narrow, the space of Nuel (NS) wider. Four rows of outer hair cells (1, 2, 3, 4) and one row of inner hair cells (IHC) are clearly discernible (× 525). (B) Half a coil (11 mm) from the basal end, the tunnel and the space of Nuel have both widened considerably. The fluid spaces around the outer hair cells are also partially developed. The tectorial membrane has not yet become completely separated from the inner ridge (IR). The mesenchymal layer under the basilar membrane is thick and contains two spiral vessels (× 525).

but instead is due to the real growth of the supporting cells and the concomitant expansion of the space by the secretion of fluid (Van der Strict, 1919; Weibel, 1957). At this time, the cytoplasm in the inner and the outer pillar cells is richly vacuolated (Figure 8).

The tunnel of Corti appears later than the space of Nuel. In the human fetus, it has become discernible in the basal coil at 15 weeks of age (Bast & Anson, 1949). During the succeeding month, the development is gradually extending toward the apex (Figure 7).

At 6 months (32 cm), the surface pattern of the organ of Corti has changed considerably. The whole surface area has widened, increasing the distance between the rows of hair cells. Particularly striking is the elongation of the inner pillar heads and the consequently greater distance between the inner hair cells and the first row of outer hair cells (Figure 8).

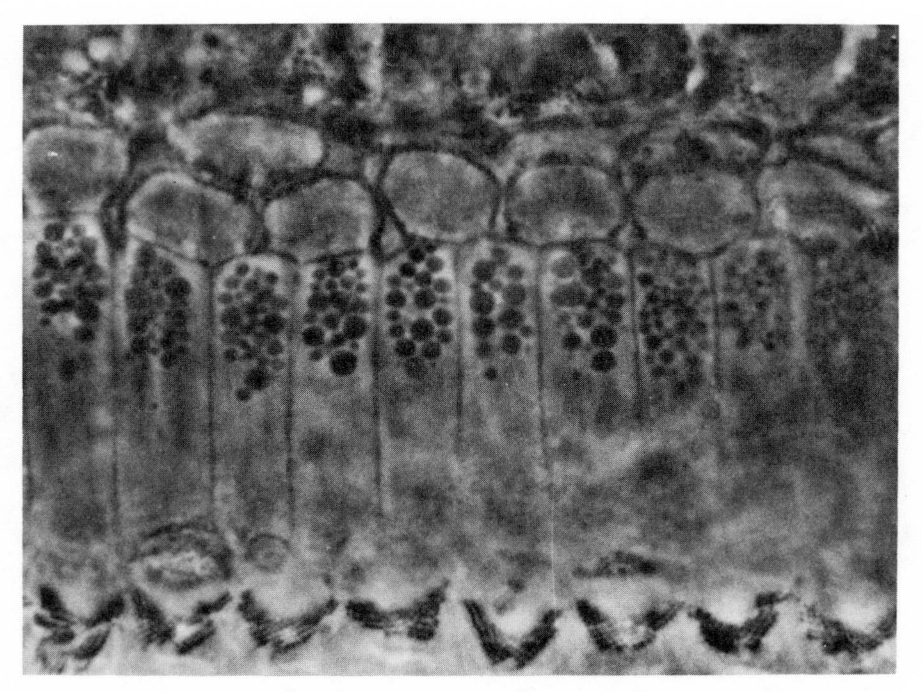

Figure 8. Organ of Corti of a 6-month human fetus; surface preparation from the upper basal coil (17 mm) showing the large distance between the inner hair cells (IHC) and the first row of outer hair cells as compared with that of one month earlier (cf. Figure 6). Note the granules in the inner pillar heads. One supernumerary inner hair cell is seen above the regular row of cells. The hairs of the first row of outer hair cells (1) are in focus (× 1570).

A second striking change in the pattern of sensory and supporting cells is that irregularities have become more common and more conspicuous than at earlier stages. Nevertheless, the regular basic pattern is retained, especially in the basal end, as the irregularities tend to increase progressively toward the apical end (Figure 9).

Some types of irregularities in the human organ of Corti occurred only infrequently, giving rise to the question of whether these were to be regarded as normal variants or as malformations. One such example is to be found in a linear deviation of the pattern of outer hair cells, which was first observed in one fetus. As the study continued, and as interest became focused on the typing of various irregularities, the same pattern was observed frequently in fetuses as well as in adults. A similar deviation was also seen in one squirrel monkey (Figure 10). Such observations tend to take this particular irregularity out of the category of malformation.

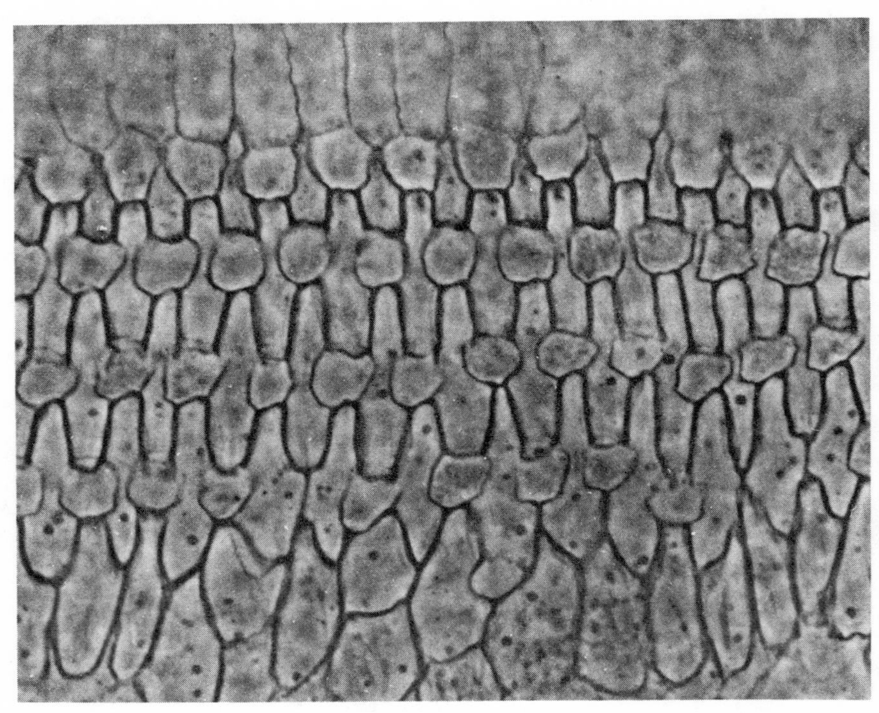

Figure 9. Organ of Corti from a premature child (fetal age, 8 months), surface specimen from upper middle coil (26 mm), showing irregularity of the pattern of outer hair cells. The general arrangement of the cells in three or four rows is present. The positions of a few lost cells can be clearly seen (× 890).

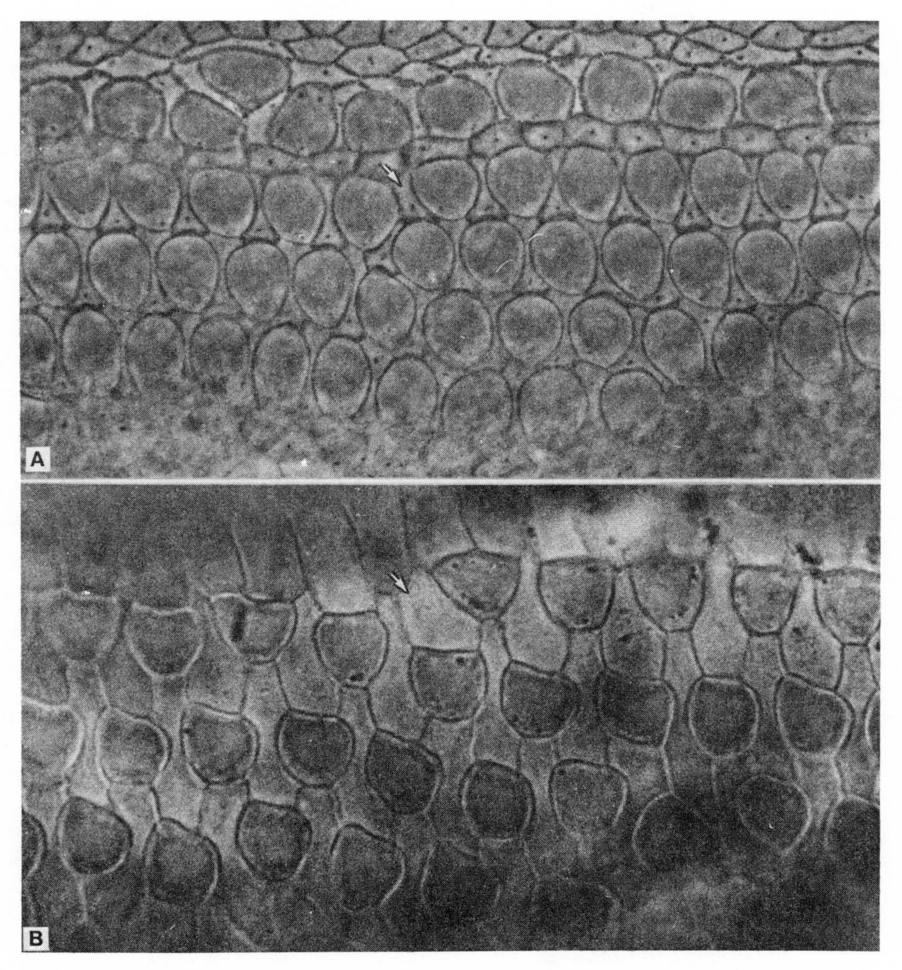

Figure 10. Surface preparation showing an interesting variety of the normal pattern. (A) 4-month human fetus, lower basal coil. Note the deviation in the linear configuration at the midfield (arrow) (× 670). (B) Adult squirrel monkey, one coil from the basal end, showing a similar arrangement of the linear configuration (× 1000).

At the fetal age of 25 weeks, the cochlea has attained its final size and is fully enclosed in the bony capsule. At this stage the organ of Corti resembles that of the adult (Bast & Anson, 1949; Ormerod, 1960), with the exception that no fluid spaces are as yet evident in the apical coil (Bredberg, 1968).

The limbus spiralis and the inner spiral sulcus are formed from the inner ridge. The outer spiral sulcus is formed from the outer part of the outer ridge.

The tectorial membrane appears as a jellylike deposit on the inner ridge. Later it is separated from the cell surface and is located between the limbus and the inner hair cells. According to Weibel (1957), the fluid under the tectorial membrane is secreted by the inner sulcus cells, which are richly vacuolated at this time.

At 8 months the pattern of sensory and supporting cells shows the essential characteristics of the adult cochlea. The fluid spaces are well developed throughout the organ of Corti (Figure 10).

In two out of four subjects studied at this age, a certain degeneration of outer hair cells was observed. It must be emphasized that the cochleas were obtained from stillborn, asphyxiated, premature children. All stages of degeneration of the hair cells were observed, from the early "collapse figures" to the "phalangeal scars." The loss of sensory cells was spread diffusely throughout the cochlea. In most places single sensory cells, rather than groups of cells, were missing. The extent of degeneration was less than 5%.

Innervation of the Organ of Corti

The development of the innervation of the organ of Corti has been most thoroughly studied by Retzius (1894), Cajal (1919), Lorente de No (1926), and Tello (1931). At a very early age, even before the two epithelial ridges appear, a subepithelial plexus of nerves can be discerned. Lorente de No demonstrated in the rat fetus that the nerves have established connections with the sensory cells by the time these can be first identified in a section (14-mm rat fetus, basal coil). He expressed the opinion that the nerve fibers and the sensory cells develop independently up to a certain point but are thereafter interdependent for the completion of their development. Nerves and sensory cells establish contact with each other at about the same time as the perilymphatic spaces begin to be formed (Lorente de No, 1926).

The myelination of the nerves occurs much later. Bechterew (1885) found that the myelination of the cochlear nerve in the human fetus takes place at a fetal length of 30 cm. Lorente de No (1926) indicated that in the rat most of the nerves acquired their myelin sheath on the fifth day after birth.

It is important in the study of development and differentiation of the organ of Corti to know when the ear starts to function as a hearing organ.

On the basis of morphological criteria, it has been assumed that responsiveness to acoustic stimulation would appear at the fetal age of 6 months (Bast & Anson, 1949; Ormerod, 1960; Wedenberg, 1965); however, in animal experiments, it has been shown that the cochlea may begin to react electrophysiologically before it is entirely mature morphologically. Thus, the fluid spaces were found not to be fully developed before response to sound stimulation could be elicited (Larsell *et al.*, 1944; Mikaelian & Ruben, 1965). Kikuchi and

Hilding (1965) found that the efferent nerves and nerve endings of the mouse were present several days before fluid spaces were formed. The appearance of the efferent nerve endings, on the other hand, was delayed until it had been possible to elicit microphonics and action potentials from the eighth nerve (Mikaelian & Ruben, 1965). The animal experiments thus suggest the possibility that the human fetal cochlea may begin to react to sound stimulation before the age of 6 months.

A few observations of auditory function in human fetuses have been reported. Fleischer (1955) noted movements of the 7- to 9-month-old fetus following a tonal stimulus delivered some distance from the abdomen of the pregnant woman. The stimulus intensity was 115 dB and the frequencies that induced reactions were 500 Hz and 1000 Hz. Murphy and Smyth (1962) and Johansson, Wedenberg, and Westin (1964) noted, with the aid of ECG, a rise in the fetal heart-rate following tonal stimulation. In fetuses 2 to 7 weeks before term, Johansson *et al.* (1964) observed this reaction in response to a sound stimulus (3000 Hz, 110 dB) delivered on the abdomen of the pregnant woman. Wedenberg (1965) reported that such reaction had been recorded in the 26th week of pregnancy.

The question of functional maturation of the cochlea has important clinical

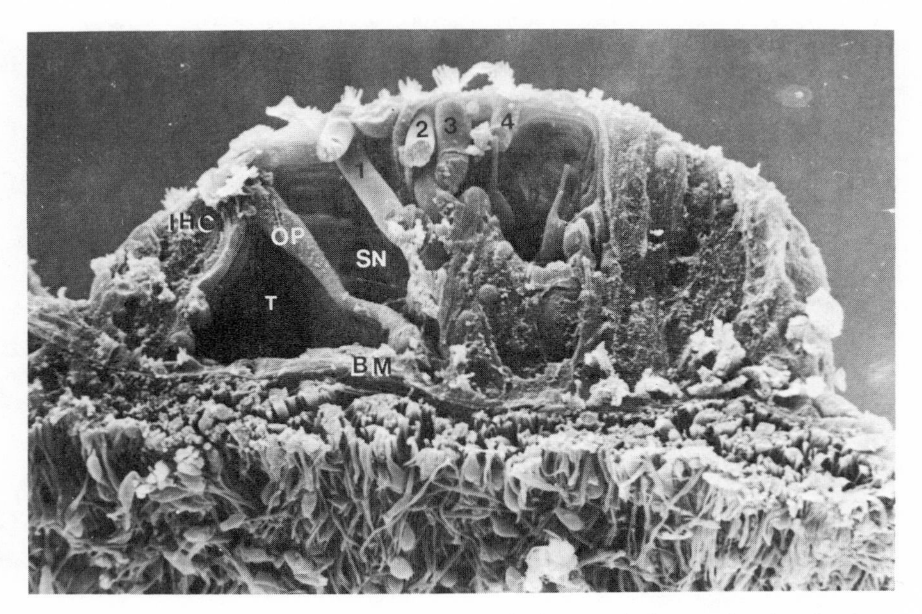

Figure 11. Surface view of the adult human organ of Corti as seen in the scanning electron microscope. On the ridge of the organ are seen three to four rows of outer hair cells and one row of inner hair cells (× 300).

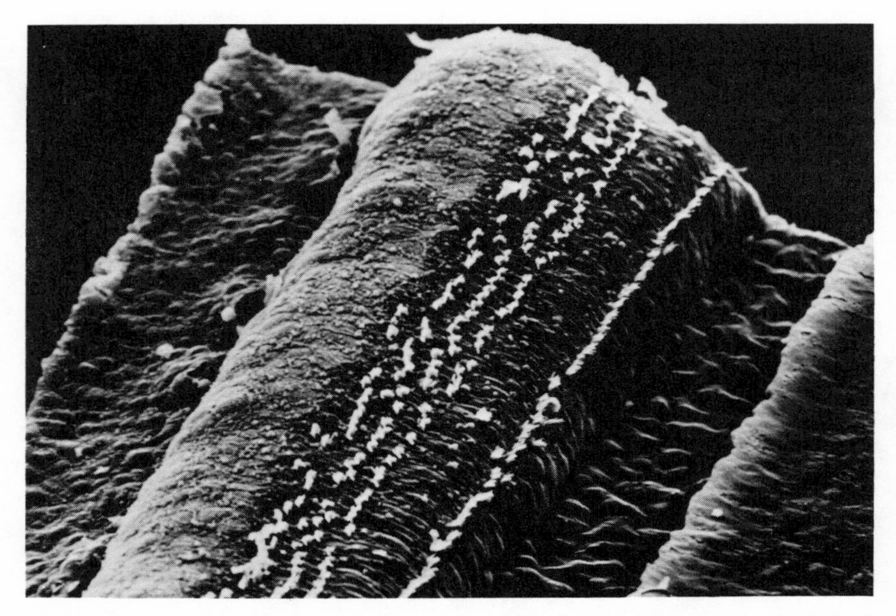

Figure 12. Cross section through the adult human organ of Corti showing an inner hair cell (IHC), the tunnel of Corti (T), the space of Nuel (SN), and four rows of outer hair cells (1, 2, 3, 4). The organ of Corti rests on the basilar membrane (BM). Outer pillar cells, OP. Scanning electron microscopy (× 590).

implications relating to the criteria for therapeutic abortion. Several countries now permit operations when there is compelling reason to believe that the mother might give birth to a child with some serious developmental defect. The early diagnosis of dysfunction in the organ of Corti and the implicit probability of cochlear damage might contribute significantly to the justification of a therapeutic abortion as, for example, in cases of maternal rubella. If the cochlea is capable of functioning earlier than the 24th week, it might be possible to develop hearing tests that could be made at such time.

Figures 11 and 12 show the anatomy of the adult human organ of Corti as seen in the scanning electron microscope.

Implications for Auditory Development

The available anatomical and electrophysiological evidence suggests that the inner ear begins to function around the 5th to 6th month of fetal life.

Anatomical and electrophysiological studies in animals suggest further that the cochlea begins to respond to sounds first in the midfrequency range with sensitivity to higher and lower frequencies appearing somewhat later in development. Both the anatomy and innervation of the organ of Corti are complete at birth. Hence developmental changes in audition following birth are probably not attributable to further maturation of the mechanisms of the inner ear.

The structure of the middle ear, however, matures more slowly. Although the ossicles appear to have their final shape and size at birth, the size of the middle-ear cavity continues to change into early childhood. Since cavity size influences sound transmission, it is possible that some changes in responsivity to sound are due to maturation or growth of the structures of the middle ear.

Finally, changes in the size and shape of the external ear and ear canal can be expected to affect, at the very least, auditory sensitivity in the infant and in the child.

References

Alexander, G. Entwichlungsgeschichte, Anthropologie, Varietaten. In A. Denker & O. Kahler (Eds.), *Handbuch der Hals- Nasen- Ohren- Heilkunde,* (Vol. 6). Berlin: Springer Verlag, 1926.

Änggård, L. An electrophysiological study of the development of cochlear functions in the rabbit. *Acta Oto-laryngologica.* (Stockholm), *Supplementum 203,* 1965, 1–64.

Anson, B. J. The early development of the membranous labyrinth in mammalian embryos, with special reference to the endolymphatic duct and utriculo-endolymphatic duct. *Anatomical Record,* 1934, *58,* 127–137.

Bast, T., & Anson, B. J. *The temporal bone and the ear*. Springfield, Ill.: Charles C Thomas, 1949.

Bechterew, W. Uber die innere Abteilung des Strichkorpers und des achten Hirnnerven. *Neurologisches Centralblatt,* 1885, *3,* 145–147.

Bredberg, G. Cellular pattern and nerve supply of the human organ of Corti. *Acta Oto-laryngologica.* (Stockholm), *Supplementum 236,* 1968, 1–135.

Bredberg, G., Ades, H. W., & Engstrom, H. Scanning electron microscopy of the normal and pathologically altered organ of Corti. *Acta Oto-laryngologica.* (Stockholm), *Supplementum 301,* 1972, 3–48.

Cajal, S. Ramon y. Accion neurotropica de los epitelios. *Trabajos des Instituto Cajal de Investigaciones Biologicas,* 1919, *17,* 181. Also in Cajal, S. Ramon y: Studies on vertebrate neurogenesis. The mechanism of development of intra-epithelial sensory and special sense nerve terminations. Springfield, Ill.: Charles C Thomas, 1960.

Crowley, D. E., & Hepp-Reymond, M. -C. Development of cochlear function in the ear of the infant rat. *Journal of Comparative Physiological Psychology,* 1966, *62,* 427–432.

Davis, H. *Hearing and deafness: A guide for laymen*. New York: Murray Hill, 1947.

Fleischer, K. Untersuchungen zur Entwicklung der Innenohrfunktion (intrauterine Kindsbewegungen nach Schallreizen), Ztschr. *Laryngologie, Rhinologie, Otologie,* 1955, *34,* 733–740.

Held, H. Die cochlea der Sauger und der Vogel, ihre Entwicklung und ihr Bau. In A. Bethe (Ed.), *Handbuch der normalen und pathologischen Physiologie*. (Vol. 11). Rezeptionsorgane, 1926, pp. 467–541.

Igarashi, Y., & Ishii, T. Embryonic development of the human organ of Corti. Electron microscopic study. *International Journal of Pediatric Otorhinolaryngology,* 1980, *2,* 51–62.

Johansson, B., Wedenberg, E., & Westin, B. Measurement of tone response by the human foetus. A preliminary report. *Acta Oto-laryngologica* (Stockholm), 1964, *57,* 188–192.

Kikuchi, K., & Hilding, D. The development of the organ of Corti in the mouse. *Acta Oto-laryngologica* (Stockholm), 1965, *60,* 207–222.

Kolmer, W. Gehörorgan. In W. Möllendorff (Ed.), *Handbuch der mikroskopischen Anatomie des Menschen* (Vol. 3). Berlin: Springer Verlag, 1927.

Larsell, O., McCrady, E., & Larsell, J. F. The development of the organ of Corti in relation to the inception of hearing. *Archives of Otolaryngology* (Chicago), 1944, *40,* 233–248.

Lorente de No, R. Etudes sur l'anatomie et la physiologie du labyrinthe de l'oreille et du VIIIe nerf. *Trabajos del Institute Cajal de Investigaciones Biologicas,* 1926, *24,* 53–153.

Lorente de No, R. Anatomy of the eighth nerve. Central projections of the nerve endings of the internal ear. *Laryngoscope,* 1933, *43,* 1–38.

Mikaelian, D., & Ruben, R. J. Development of hearing in the normal CBA-J mouse. Correlation of physiological observations with behavioural responses and with cochlear anatomy. *Acta Oto-laryngologica* (Stockholm), 1965, *59,* 451–461.

Murphy, K. P., & Smyth, C. N. Response of foetus to auditory stimulation. *Lancet,* 1962, *1,* 972–973.

Ormerod, F. C. The pathology of congenital deafness. *Journal of Laryngology and Otology,* 1960, *74,* 919–950.

Rasmussen, A. T. *Outline of neuroanatomy.* Dubuque, Iowa: William C. Brown, 1943.

Retzius, G. *Das Gehörorgan der Wirbelthiere. II. Das Gehörorgan der Reptilien, der Vogel und der Saugethiere.* Stockholm: Samson & Wallin, 1884.

Ruben, J. Development of the inner ear of the mouse. A radioautographic study of terminal mitoses. *Acta Oto-laryngologica.* (Stockholm), *Supplementum 220,* 1967, 1–44.

Tello, J. F. Le réticule des cellules ciliées du labrinthe chez la souris et son indépendance des terminaisons nerveuses de la VIlle paire. *Trabajos del Instituto Cajal de Investigaciones Biologicas,* 1931, *27,* 151–186.

Van der Stricht, O. The development of the pillar cells, tunnel space, and Nuel's spaces in the organ of Corti. *Journal of Comparative Neurology,* 1919, *30,* 283–321.

Van der Stricht, O. The arrangement and structure of sustentacular cells and hair-cells in the developing organs of Corti. *Contributions of Embryology,* 1920, *9,* 109–142.

Wada, T. Anatomical and physiological studies on the growth of the inner ear of the albino rat. *Wistlar Institute of Anatomy and Biology,* 1923, Memoirs No. 10.

Wedenberg, E. Prenatal tests of hearing. *Acta Oto-laryngologica.* (Stockholm), *Supplementum 206,* 1965, 27–32.

Weibel, E. R. Zur Kenntnis der Differenzierungsvorgange im Epithel des Ductus cochlearis. *Acta Anatomica* (Basel), 1957, *29,* 53–90.

Wersäll, J., & Flock, A. Morphological aspects of cochlear hair-cell physiology. In *Henry Ford Hospital International Symposium.* Boston: Little, Brown, 1967.

CHAPTER 2

Physiology of the Developing Auditory System

J. J. Eggermont

Department of Medical Physics and Biophysics
University of Nijmegen
Nijmegen, The Netherlands

Introduction

The auditory system as a whole can only be studied by behavioral tests. When we speak of hearing, we are considering something that leads to a motor response, either an orientation toward the sound source or an answer to it. Hearing involves localization and identification of the sound. Electrophysiological methods can never describe the system completely, since it is impossible to determine with these methods whether a sound was actually heard. Thus, an electrophysiologically normal system is a necessary condition for normal hearing, but it is not a sufficient condition. Depending on the recording site, one can only test the system up to that point. By testing various stations along the auditory pathway, one may obtain time courses of development at all these points. This allows one, for example, to distinguish between maturation at the cochlear level and maturation of the central nervous system.

In the mathematical analysis of biological systems, one often finds variables that are decreasing or increasing in proportion to their magnitude and that lead, respectively, to exponential decay and growth. Changes like this might be important for the developing auditory system as well. In this chapter, a model of auditory development is presented which presumes, as its first property, that all changes can be adequately described by a combination of exponential changes.

When recording from higher parts of the auditory system, one generally accepts that the changes found during development will never proceed faster than those at more peripheral levels (i.e., never faster than the maturation of the cochlear receptor). An extreme version of this hypothesis would be that the developing auditory system reflects the developing ear. A slightly weaker version of this hypothesis forms the second part of our model of development. We will assume that most changes in the auditory system can be understood on the basis of the development of the inner ear.

Of course, it is generally accepted that maturational processes in the central nervous system (CNS) will also play a role. Among these are the myelination process, the gradually increasing efficiency of synapses, and the onset of convergent and divergent neural connections that lead to more straightforward connections and possibly to the elimination of synapses (Purves & Lichtman, 1980). If these central mechanisms have time constants that differ by at least a factor of 2 from those describing the changes in the peripheral process, we should be able to statistically detect their presence in the electrophysiological data. In this case one could postulate a sum of two exponential functions to fit the data. A note of caution, however, is required when adding more exponential functions together to fit the data since each added exponential will continue to provide a better fit (in the sense of the mean square error) to the data (see Riggs, 1970).

Our interest then will be focused on two basic questions: When is the human auditory system mature? What electrophysiological measures reflect this? Our search for the onset of development relates to these questions with the hope that they will be good indicators of prolonged immaturity or retardation. From a clinical point of view, we need to determine good diagnostic measures of retardation in sensory development. It appears, therefore, that the study of the developing auditory system might provide a basis for the study of retardation in auditory development.

Studies on the developing auditory system, and especially its electrophysiology, have been executed on a large number of bird and mammalian species (chickens and ducklings are used in most studies because of the ease with which they can be studied in the prehatched stages). This review is restricted primarily to mammals, in particular, to cats and humans. Occasionally, evidence from studies on mice, rats, and guinea pigs is considered. It will become apparent that maturation proceeds along general lines in different species and that, in humans especially, the various maturational processes are more clearly separated than in lower mammals.

There have been several reviews of the literature on the development or ontogeny of the auditory system, including recent reviews by Javel (1980) and Ruben and Rapin (1980). One particular scholarly review by Rubel (1978) strongly puts emphasis on the development of hearing in birds but also provides a complete account of anatomical, electrophysiological, and behavioral studies with mammals. The present chapter will, however, have a shift in emphasis. First, the role of cochlear maturation and the way it is reflected along the neuraxis will be discussed. Second, there will be a search for time constants that characterize postulated exponential changes in a large number of response parameters. A considerable amount of attention will be given to maturational changes in the human auditory system that have recently come to light from the application of brainstem (auditory) potentials.

The chapter consists of three parts. The first part presents a compilation of data from several authors on various levels in the auditory system of the cat and covers most of the parameters that have been studied. The second part describes characteristic changes in the human brainstem auditory evoked potential (BAEP) and again shows the convergence of findings from various laboratories. The third part speculates on the extrapolation of results obtained from the cat (and currently unavailable in humans) to the hearing processes of the human neonate. Finally, since all these sections involve detailed descriptions of, and extrapolations from, several different electrophysiological measures, it is suggested that the reader who is not familiar with these techniques consult Yost and Nielson (1977, Chapter 6, pp. 66–85, and Chapter 7, pp. 104–106), where a brief and lucid description of these measures is provided.

Maturation in the Cat

Threshold

In the cat, the cochlear microphonic (CM) is detectable from birth, but it takes a few hours after birth before the compound action-potential (AP) of the auditory nerve becomes detectable. CM amplitude for a 1-kHz stimulus at 80 dB SPL changes from 80 μV at Day 6 to an adult value of 270 μV at Day 15. AP parameters, in general, take about a month to reach maturity (Romand, 1971). When input–output curves are constructed for the first negative peak (N_1) of the compound action potential, the resulting function relating peak N_1 amplitude to sound intensity rises steeply in 2-day-old kittens. Around the end of the first week, an inflection appears in the curve and, at Day 13, the input–output curve shows a shallow part that extends over a 10-dB range before accelerating sharply for higher intensities. Finally, at Day 30, the shallow part covers 30 dB as in the adult (Carlier & Pujol, 1978).

These changes in the slope of the input–output curve for N_1 are paralleled by a decrease in threshold value for N_1. When the N_1 threshold is considered as a function of tone burst frequency (an N_1 audiogram), it is observed that, around the 15th day after birth, functional development is greater in the lower frequency region where maturity is reached at around 36 to 40 days. At the high-frequency end of the cochlea, however, maturation only appears to be complete after around 50 days (Moore & Irvine, 1979).

In combining threshold data obtained from various investigations using different physiological measures (Aitkin & Moore, 1975; Brugge, Javel, & Kitzes, 1978; Moore & Irvine, 1979), one observes (Figure 1) that the difference follows an exponential decrease to adult threshold. We have plotted the data with reference to the deviation from adult values for each of the separate measures, since the N_1 thresholds (50 μV criterion) are elevated about 40 dB relative to comparable single-unit thresholds (Moore & Irvine, 1979). For the Aitkin and Moore (1975) and Moore and Irvine (1979) studies, thresholds are plotted separately for low ($\leq$ 1 kHz) and high frequencies. The data from Brugge *et al.* (1978) are from single units in the anteroventral cochlear nucleus, the Aitkin and Moore (1975) results are from single units in the inferior colliculus, whereas the Moore and Irving data are N_1 thresholds from the AP of the auditory nerve. An examination of Figure 1 indicates that low- and high-frequency thresholds mature at comparable rates. In addition, the time course of the threshold changes seems to be the same for the N_1 component of the AP recorded from the auditory nerve, as it is for single units in the cochlear nucleus and the inferior colliculus.

Thus, the difference from adult threshold values changes with age according to an exponential law:

$$Th(t) - Th(a) = 98.5 \exp(-t/12.2) \quad (1)$$

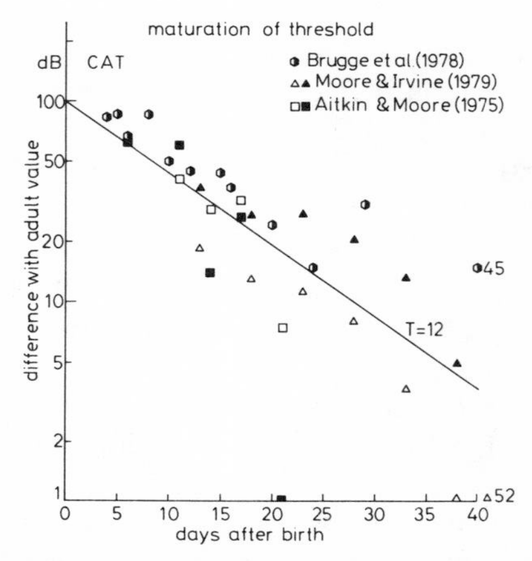

Figure 1. Maturation of threshold. Difference from adult value is plotted as a function of age on semilogarithmic coordinates. Open symbols are for frequencies $\leq$1 kHz. Solid symbols are for frequencies $\geq$ 1 kHz. Curve fit with a single exponential (time constant 12 days) has been drawn.

where *Th(a)* is the adult threshold and *Th(t)* is the threshold at time *t*. The curve provides a reasonable fit to the data ($r^2 = 0.57$) with a time constant (T_{th}) for threshold changes of approximately 12 days ($T_{th} = 12.2$). This means that after each 12 days the differences in threshold values have decreased by a factor of 2.72.

Frequency Selectivity

A property of the auditory system known to be associated with changes in threshold value is the degree of tuning of the individual auditory nerve fibers, or primarylike fibers in higher centers. In guinea pig cochleas, Robertson and Manley (1974) found a very strong correlation between degree of tuning and threshold at the characteristic frequency (Figure 2). The degree of tuning Q_{10dB} (Kiang, Watanabe, Thomas, & Clark, 1965) is often measured by dividing the bandwidth of the tuning curve at 10 dB above the characteristic frequency threshold into the value of the characteristic frequency. Thus, narrow bandwidths (sharp tuning) yield high Q_{10dB} values. Robertson and Manley's data show that

$$Q_{10dB} = 9 - 0.09\ Th\ (CF) \qquad (2)$$

in which *Th(CF)* indicates the threshold in decibel value at the characteristic frequency of the neuron. This holds for the range of characteristic frequencies

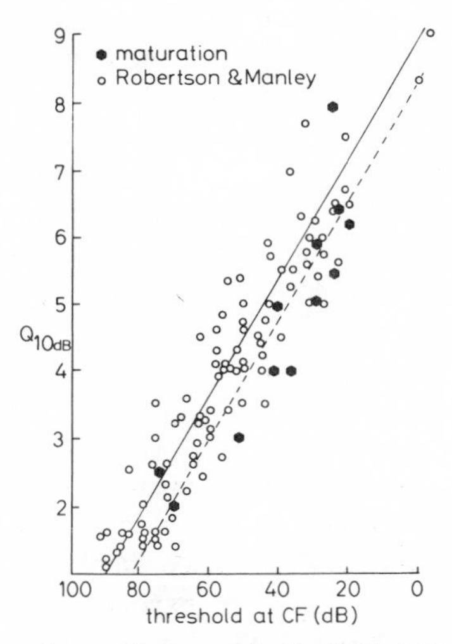

Figure 2. Comparison between Q_{10dB} changes with threshold at the characteristic frequency for maturing cochleas' and "normal" adult cochleas'. Solid line: regression for the Robertson and Manley (1974) data. Dashed line: regression for the maturation data (Saunders *et al.*, 1980).

between 12 and 19 kHz in the spiral ganglion of the guinea pig. From this equation, we note that for units having low thresholds (near 0 dB), the expected Q_{10dB} will be around 9 and that in neurons with thresholds around 90 dB SPL, the tuning will be very poor ($Q_{10dB} \simeq 1$).

Having observed the gradual improvement of threshold values during maturation (for N_1 as well as for single units), one wonders if these changes are also paralleled by changes in the Q_{10dB} value. For the cat, changes in Q_{10dB} have been observed for neurons in the inferior colliculus (Aitkin & Moore, 1975) showing that in the kitten (14- to 28-days-old) the majority of Q_{10dB} values were in the range of 0 to 3, while in adult cats they ranged from 3 to 6. As in normal adult animals and also for the kitten, the fibers with higher characteristic frequencies were found to have larger Q_{10dB} values. Corroborating evidence comes from aging mouse pups where this general trend was also found (Shnerson & Willott, 1979; Willott & Shnerson, 1978) for inferior colliculus units.

Tuning curves for single auditory nerve fibers in the kitten show that for characteristic frequency values below 1.5 kHz the average $Q_{10dB} = 0.9 \pm 0.3$ at age 20 days, while at 30 days the Q_{10dB} increases to 2.3 ± 0.7. In contrast, at characteristic frequencies above 1.5 kHz, the Q_{10dB} values change only from 4.5 ± 3.2 to 4.7 ± 2.4 (Romand, 1979). Thus, although the threshold data of Irvine and Moore (1979) suggest prolonged maturation, especially in the high-frequency part of the cochlea, these tuning-curve data mainly suggest maturation in the low-frequency region.

Indications about the tuning of single nerve fibers can also be obtained from AP or other evoked potentials by tone-on-tone masking procedures (Dallos & Cheatham, 1976; Eggermont, 1977). This method has been applied to the kitten with basically the same results. At the higher frequencies (e.g., 8 kHz), the Q_{10dB} changes from around 1 at Day 11, and 4 at Day 13, to a nearly adult value of 7 at Day 16. Thresholds in this period change from 80 dB SPL down to 0 dB (Carlier, Lenoir, & Pujol, 1979).

Saunders, Dolgin, and Lowry (1980) studied the development of tuning in the cochlear nucleus of mice on the basis of evoked-potential tuning curves. These data are sufficiently detailed to permit exponential modeling:

$$Q(t) = Q(a)[1 - B \exp(-t/7)], \; r^2 = .70 \qquad (3)$$

where $Q(t)$ is the degree of tuning at time t, $Q(a)$ is the adult value, and B is a constant. $Q(a)$ and B are dependent on the characteristic frequency (CF) (in this case the test-tone frequency used), but for low as well as high frequencies it appeared that an exponential fit with a time constant $T_Q = 7$ days could be made.

Thus, although thresholds decrease exponentially, the degree of tuning approaches an adult value following a $[1 - \exp(-t/T)]$ law. In order to compare this with the relation found between Q_{10dB} and threshold at CF for the single-unit data in the guinea pig, we calculated from Saunders *et al.* (1980) that for 16 and 20 kHz the following relation (see Figure 2, filled hexagons) describes the data sufficiently well ($r^2 = 0.80$):

$$Q_{10dB} = 8.3 - 0.09\ Th(CF) \tag{4}$$

This shows a striking parallel with data from Robertson and Manley (1974) as compiled in equation (2) and shown in Figure 2.

Firing Rate and Poststimulus-Time Histogram (PSTH)

Another feature of maturation that covaries with threshold is the saturation firing rate (Brugge *et al.*, 1978) of single units. The average adult firing rate of about 305 spikes per sec is reached again in an exponential way:

$$R(t) = R(a)\ [1 - 3.03\ \exp(-t/5.56)],\ r^2 = 0.73 \tag{5}$$

where $R(t)$ is the number of spikes per sec at time t and $R(a) = 305\ s^{-1}$ is the adult value. The estimated time constant, T_R, is about 5.6 days.

The results of Brugge *et al.* (1978) are also demonstrated by Romand and Marty (1975) who observed changes in the firing rate from (on average) 40 spikes per sec at 9 to 10 days, to 132 spikes per sec at 30 days, with adult values at 232 spikes per sec. In contrast to Brugge *et al.* (1978), Romand and Marty (1975) do not find adult levels at 30 days.

Part of these changes in firing rate may be caused by changes in the way the neural units fire to short tone bursts (Carlier, Abonnenc, & Pujol, 1975). They observed in the auditory nerve of the cat changes from an "on-type" post stimulus-time histogram (PSTH) to a rhythmic response and finally, a continuous response pattern.

Changes in the PSTH pattern have also been reported for the cochlear nucleus where, at birth, exclusively rhythmic PSTHs are found and, at 9 to 10 days, sustained responses replace the rhythmic ones (Romand & Marty, 1975). For the inferior colliculus of the cat, the unit responses from Day 2 onward also represent these aspects: "on" responses, rhythmic responses, and responses with intermittent silence. These types of response change gradually and disappear completely around 14 days after birth, after which adultlike continuous responses are recorded (Romand, Granier, & Marty, 1973). Moore and Irvine (1980) also find continuous responses at 21 to 30 days and so-called onset, pauser, and burst-type responses were observed in animals of all ages. Pujol (1969) reported that continuous responses only appeared at the end of the first month. A schematic summary of these findings is given in Figure 3.

Latency

As threshold decreases and saturation firing rate increases, one observes that the latency of the first spike in response to clicks or tone bursts decreases. The same holds true for the latency of the N_1 or other evoked potentials recorded from within various nuclei. Figure 4 presents a compilation of changes in mean latency values reported by various authors for the auditory nerve (Carlier *et al.*, 1975; Romand, 1971), the cochlear nucleus (Romand & Marty, 1975), the olivary complex (Romand *et al.*, 1973), the inferior colliculus, and the cortex (Pujol, 1969).

The time course of the changes in all cases suggests a single exponential decrease. The observations for cochlear nucleus and olivary complex have been taken together, whereas the single-unit data for the auditory nerve and the AP data have been treated separately. Exponential curve fits resulted in:

$$\begin{array}{ll} \text{N VIII, single unit} & L(t) - L(a) = 27.8 \exp(-t/6.25) \\ \text{N VIII, AP} & L(t) - L(a) = 2.6 \exp(-t/7.7) \\ \text{Cochlear nucleus} & \\ \text{Olivary complex} & L(t) - L(a) = 12.2 \exp(-t/5.9) \\ \text{Inferior colliculus} & L(t) - L(a) = 35.4 \exp(-t/6.25) \\ \text{Cortex} & L(t) - L(a) = 83.4 \exp(-t/5.6) \end{array} \qquad (6)$$

where $L(t)$ is the latency at time t, and $L(a)$ is the adult latency.

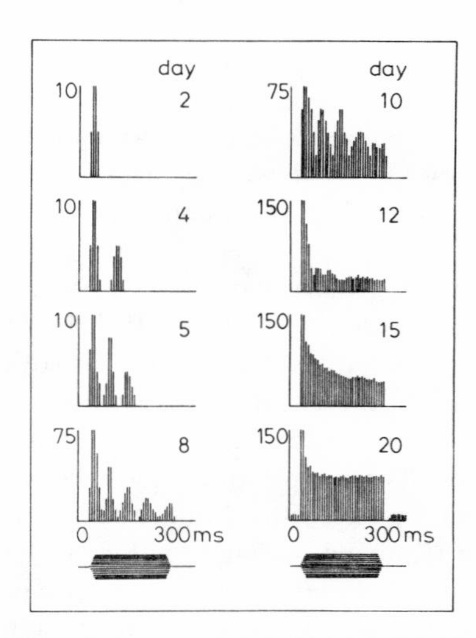

Figure 3. Postnatal development of tone burst PSTH in the auditory nerve. Initial onset responses change after 4–5 days into rhythmic (chopper type) responses. At 12–15 days sustained responses (primary like) appear. Importance of the transient character during first 50 ms gradually decreases. (After Carlier *et al.*, 1975.)

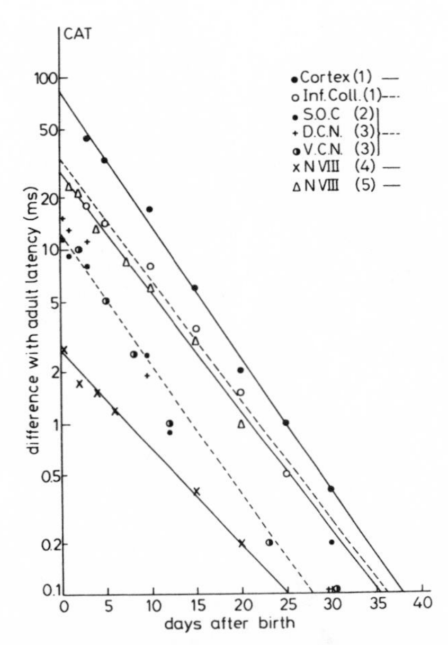

Figure 4. Latency changes from auditory nerve to cortex as a function of age. Differences from adult latency are plotted on semilogarithmic coordinates and exponential functions have been fitted to the data points. Original latency data obtained from (1) Pujol (1969), (2) Romand *et al.* (1973), (3) Romand and Marty (1975), (4) Romand (1971), and (5) Carlier *et al.* (1975).

The set of time constants ranging from 5.6 to 7.7 strongly suggests that the maturation of latency from auditory nerve to cortex can be adequately described by a single time constant $T_L = 6$ days.

In contrast with these results from intracranially recorded data, Shipley, Buchwald, Norman, and Guthrie (1980) indicated that for brainstem auditory evoked potentials (BAEP) the latency changes for different waves may have different time constants. By exponential curve fits, they arrived at a fastest maturing latency for Wave 2*a* ($T = 7.35$ days) and a slower process for Wave 5 ($T = 13.33$ days). Recent data by Walsh, McGee, and Javel (1981), which show the day-by-day changes in latency for Waves I and V in the brainstem auditory evoked potentials for a group of 25 kittens, allow further elaboration on the potential differences between depth recordings and surface recordings. Although Walsh *et al.* suggest a linear change in latency for Waves I and V during the first 20 days followed by an exponential change toward adult values, a single exponential fits the data equally well (Figure 5).

For Wave I their data can be represented by

$$L_I(t) - L_I(a) = 2.52 \exp(-t/12.87),\ r^2 = 0.49 \qquad (7)$$

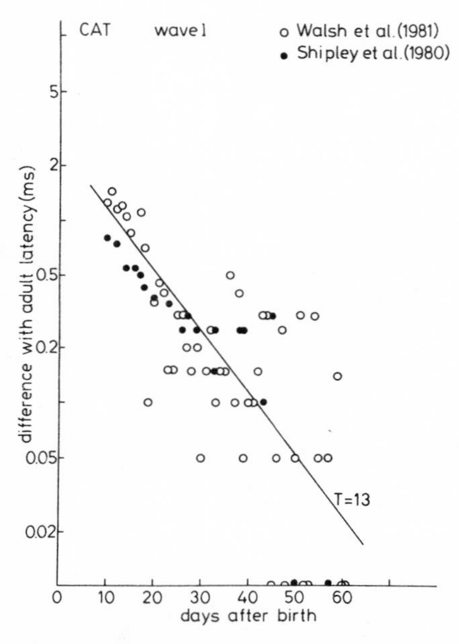

Figure 5. Latency changes for Wave I from cat BAEP as a function of age. Original data from Walsh *et al.* (1981) and Shipley *et al.* (1980) have been converted to differences from estimated adult latency. Drawn line indicates the exponential curve fit to the Walsh *et al.* data.

with an adult value $L_I(a) = 1.15$ ms. For comparison, the latency changes derived from the Shipley *et al.* data with respect to an adult value of 0.75 ms seem to compare very well with the Walsh *et al.* data. Actually, an exponential curve fit for the Shipley *et al.* data resulted in a time constant of 13.2 days, which is somewhat at variance with their own calculations.

For Wave V, the Walsh *et al.* data (Figure 6) again allow the changes to be represented by a single exponential:

$$L_V(t) - L_V(a) = 9.92 \exp(-t/12.2), \; r^2 = 0.77 \tag{8}$$

with $L_V(a) = 3.82$ ms, suggesting the *same* time constant as for Wave I. For comparison, latency data from the AVCN obtained by Brugge *et al.* (1978) have been plotted as well. Again the Shipley *et al.* data compare very well with the other results, the calculated time constant being $T = 15$ days, that is, slightly, but not significantly, longer than their own estimate. It should be noted that Wave V from Walsh *et al.* probably corresponds to Wave 4 of Shipley *et al.;* however, the *changes* in latency are still comparable.

Finally the I–V latency difference (Figure 7) for the Walsh *et al.* data again follows an exponential change:

$$L_{\mathrm{I-V}}(t) - L_{\mathrm{I-V}}(a) = 8.61 \exp(-t/11.4),\ r^2 = 0.78 \tag{9}$$

with the adult value $L_{\mathrm{I-V}}(a) = 2.68$ ms. The Shipley *et al.* data agree very well with the other and suggest a time constant of 14 days.

It therefore appears that the surface recordings indicate a larger time constant for the maturation process than the depth recordings. On average, $T = 12$ days seems to be quite representative. In contrast with the conclusions from Shipley *et al.* (1980), we do not find much evidence for a difference in time constant of the peripheral maturation process and the more central maturation process.

Latency changes in surface-recorded brainstem auditory evoked potentials, however, have been shown to be strongly linked to changes in the inferior colliculus concentration of myelin lipids (Shah, Bhargava, & McKean, 1978). Thus, one expects changes in myelination during maturation to show up in

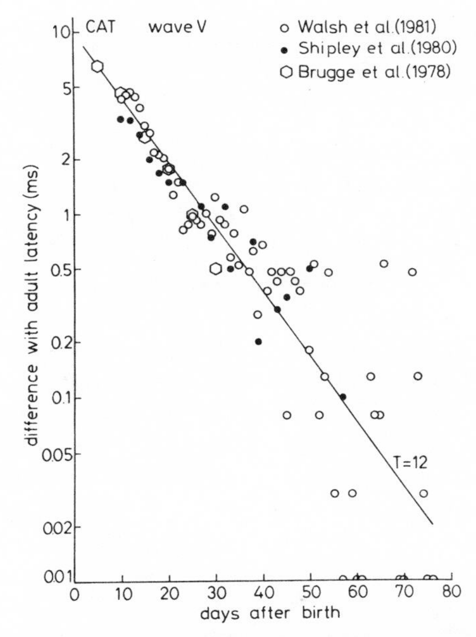

Figure 6. Exponential curve fit to changes in latency for Wave V for Walsh *et al.* data. Measurements by Shipley *et al.* (1980) are also shown. Hexagons indicate latency differences (estimated mean values) from adult values for single neurons from the anteroventral cochlear nucleus (Brugge *et al.*, 1978).

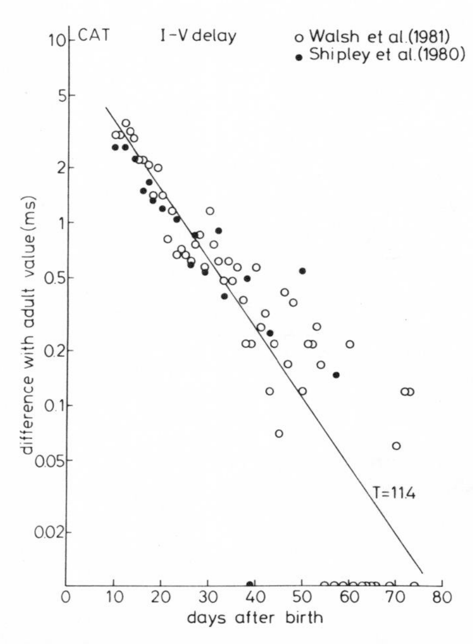

Figure 7. Exponential curve fit for I–V latency difference from adult values for Walsh *et al.* (1981) data. I–V latency difference as obtained from Shipley *et al.* (1980) are drawn in.

latency changes. When receptor maturation and changes in myelination proceed at approximately the same rate, this will not become evident.

Cochlear Mechanics

The role that is played by the maturation of the frequency analyser in the cochlea, the cochlear partition, has been investigated most thoroughly by Brugge *et al.* (1978). They established that by the end of the first postnatal week the cochlear partition may be capable of supporting a traveling wave along most of the length of the basilar membrane. It was furthermore concluded that the place–frequency relation would not differ much from the adult situation. Nevertheless, the immature cochlea shows traveling-wave and cochlear-filter time delays that are 3 to 7 ms longer than in the adult, and this disappears only near the end of the first month. This time delay may be an important factor in the latency changes observed for various places along the neuraxis.

In addition, the ability to phase-lock to the stimulus period has been investigated by the same authors. It was observed that at the end of the first postnatal week phase-lock was only evident for frequencies below 800 Hz, whereas the adult cat may show a phase preference up to 3500 Hz. This may be of importance for directional hearing capacity. In this respect, Moore and Irvine (1980) did not find evidence for postnatal development of binaural connections based on a classification of binaural response cell type. In a subsequent report, however, Moore and Irvine (1981) found some evidence for increasing sensitivity to interaural intensity differences.

Discussion

In the maturing cochlea, as well as in pathological ones, the threshold, the steepness of the input–output curve and the degree of tuning are intimately related, and the model concept of Evans (1975) describing this may well apply to the maturation process, too. The Evans model, however, links together cochlear processes; part of the threshold changes in the kitten may be caused by the changes that take place in the middle ear, which becomes air-filled around the end of the second postnatal week. From that time on, the degree of tuning becomes measurable (Aitkin & Moore, 1975). Therefore, restricting the application of the recruitment model (Evans, 1975) to the period after the first 14 days, one observes that the close parallels found during maturation between steepness of the input–output curve and Q_{10dB} warrant the hypothesis that during this developmental process the "second filter" becomes operational. Several studies relate the onset of this sharpening process to the maturation of the outer haircells and the efferent innervation thereof (Carlier & Pujol, 1978; Carlier *et al.*, 1979).

The finding of greatly elevated time delays for the traveling wave in the maturing cochlea will result in a decrease in the traveling wave velocity and therefore in the amount of synchronization of single-nerve-fiber firings to tone bursts (Eggermont, 1976). As this synchronization is reflected in the amplitude of the AP, a fixed amplitude criterion for the N_1 audiogram will incorporate this aspect (Moore & Irvine, 1979). There remains, however, some discrepancy, as Brugge *et al.* (1978) report adult-value time delays one month after birth and Moore and Irvine (1979) report prolonged maturation to around 50 days for the N_1 audiogram.

As threshold might also depend on the actual depolarization of the haircell, which, in turn, is governed by the intracellular potential and the endolymphatic potential, the investigations of Fernandez and Hinojosa (1974) become important. They observed virtually no changes in the intracellular potential during development but the endolymphatic potential increased during maturation from 8.8 ± 4.3 mV at birth to its adult value EP(a) of 80.4 ± 7.77 mV in an *S*-like fashion, well representable by

$$EP(t) = EP(a)[1 - 1.41 \exp(-t/8.56)], \quad r^2 = 0.96 \qquad (10)$$

that is, by a single exponential function with a time constant of about 8.5 days. This time constant is shorter than that describing the threshold changes. Therefore it is not likely that this effect causes the delayed mature values found by Moore and Irvine (1979) for the N_1 audiogram.

The abnormally long time delays in the kitten cochlea will be reflected in all latency changes. Additional effects on latency may arise from immature synaptic functioning. Excitatory postsynaptic potentials (EPSPs) for muscle fibers in the developing kitten have a slower rise, which reflects a slow rate of transmitter release probably caused by asynchronous arrivals at the postsynaptic membrane. This has, of course, an effect in the irregular shape of PSTHs found during the maturation process and, in addition, prolongs latency. In this way one may link together various parameter changes in the peripheral region of the auditory system. When recalling the time constants found for these changes, we must conclude that with the exception of threshold, EP, and BAEP latency changes, all parameters can be said to follow an exponential course with a time constant of around 6 days. In contrast, threshold maturation and latencies as derived from surface recordings seem to take twice as long. When comparing various recording sites, we observe that there is strong evidence from the locally generated potentials, as well as the BAEP, that the time constants applicable to the periphery of the auditory system also apply to the central part of it.

Maturation in Humans

Electrophysiological signs of development of the human auditory system are to a large extent based on surface-recorded BAEPs whereas studies of cortical evoked potential also shed light on this process. One of the characteristic features of the BAEP in humans is that the latencies of the various waves change in the same way with stimulus intensity, that is, the interwave delays are intensity independent for normal cochleas. In addition, it has been found that the site of generation along the cochlear partition has no effect on the I–V delay (Don & Eggermont, 1978; Eggermont & Don, 1980). Interwave delays are related to myelination (Shah *et al.*, 1978), and in demyelinating diseases these delays are prolonged (Shah & Salamy, 1980).

Latency of Brainstem Evoked Potentials

Extensive data on the three most prominent waves of the human BAEP as a function of the conceptional age of the neonate child are now available. From data reported by Starr, Amlie, Martin, and Sanders (1977), Salamy and McKean (1976) and Uziel, Marot, and Germain (1980), the estimated mean latency differences from adult values are shown in Figure 8, with respect to Waves I, III, and V and the I–V delay. The recordings from all three sources taken together cover an extended period from 28 weeks conceptional age until 100 weeks conceptional age (i.e., more than a year after birth). In addition, there is considerable overlap in the range studied by the different investigators. It is apparent that, except for Wave I, a fit with a single exponential is not readily feasible.

For Wave I, the results from Starr *et al.* (1977) and Uziel *et al.* (1980) can be described very well with a single exponential:

$$L_{\mathrm{I}}(t) - L_{\mathrm{I}}(a) = 451 \exp(-t/5.26) \tag{11}$$

having a time constant of about $T_{\mathrm{LI}} = 5.3$ weeks and an adult latency value, $L_{\mathrm{I}}(a) = 1.60$ ms.

However, the results from Salamy and McKean (1976) suggest very strongly a slow change of latency for Wave I after the first 45-weeks-conceptional age.

As a consequence of our basic model, the changes in Wave III and V latency might be explained by the sum of two exponentials. The curve fitting that has been made for Wave III is shown in Figure 9 and can be represented as

$$L_{\mathrm{III}}(t) - L_{\mathrm{III}}(a) = 2769 \exp(-t/4.0) + 2.62 \exp(-t/33.3) \tag{12}$$

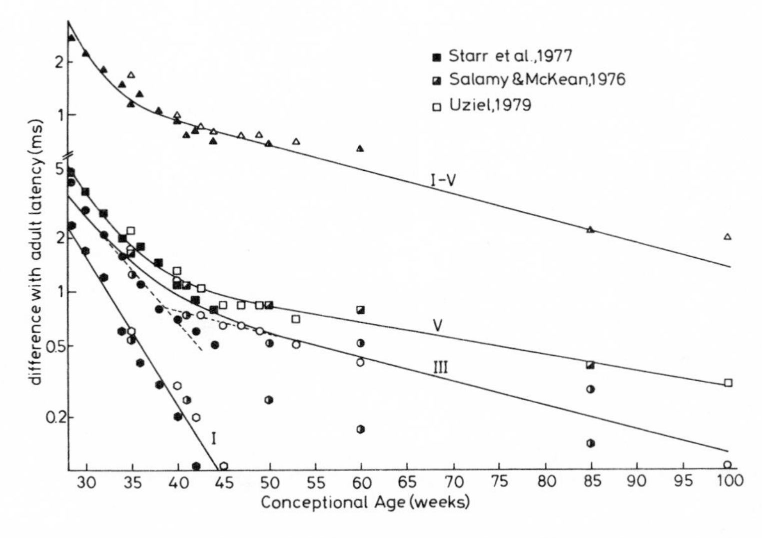

Figure 8. Latency changes for human BAEP as a function of age. For data from various sources, differences from adult latency have been compiled and plotted as a function of conceptional age. It does not seem possible to fit these data points with a single exponential curve.

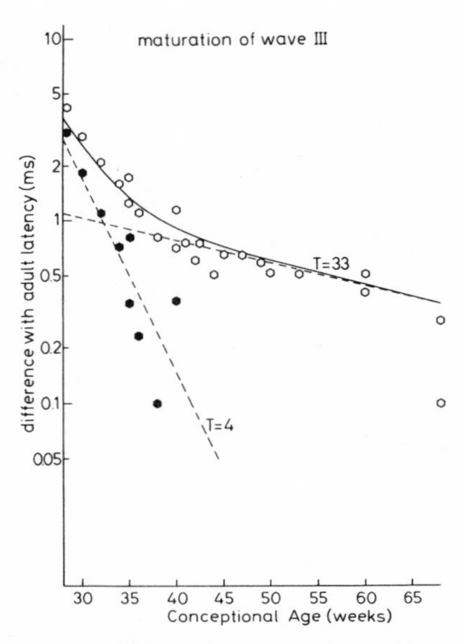

Figure 9. Fitting exponential curves to the changes in latency for Wave III (open symbols). A good fit is obtained by the sum of two exponentials having time constants of 4 and 33 weeks. Solid symbols are the same data points for the early portion of the curve after correction for the contribution of the second (and longer) exponential component. The two data points at the right are obtained at 85 and 100 weeks conceptional age.

with an adult latency value, $L_{III}(a) = 3.80$ ms. A similar procedure is applied to the Wave V data (Figure 10) resulting in

$$L_V(t) - L_V(a) = 4494 \exp(-t/4.0) + 2.22 \exp(-t/50.0) \qquad (13)$$

with an adult latency value, $L_V(a) = 5.60$ ms. It appears that the sum-of-exponentials model represents quite well the changes in the Wave III and Wave V latency. Its application results in two exponentials being sufficient, one having a time constant of 4 weeks and the other, in the range of 33 to 50 weeks.

It is tempting to assign the short time-constant process to cochlear maturation, since the value is sufficiently close to the time constant found for Wave I. The large time-constant process could then reflect the maturation of more central parts. When considering the I–V delay, one might expect that the cochlear factor will drop out, but a single exponential fit appears to be impossible. In Figure 11, the best-fitting sum of two exponentials is shown:

$$L_{I-V}(t) - L_{I-V}(a) = 50{,}008 \exp(-t/2.78) + 2.97 \exp(-t/33.3) \qquad (14)$$

with an adult latency delay, $L_{\text{I-V}} = 4.00$ ms. It appears that the longer time constant, determining the time course from 40 weeks conceptional age on, is in the range found for Waves III and V. In addition, we observe a short time-constant ($T \simeq 3$ weeks) process whose influence seems to have faded away at the normal gestational age.

For the results described above, recordings were available up to 100 weeks conceptional age. Fabiani, Sohmer, Tait, Gafni, and Kinarti (1979) report on the maturation of the "brainstem transmission time" over a much longer period than hitherto considered; their data suggest changes over nearly 5 years. These authors define their interwave period with adult values of 4.52 ms (on average) between Wave I and the vertex-negative wave following Wave V. Taking their data over the period from 1 to 12 years after birth, the changes in their interwave delay can be represented as

$$L(t) - L(a) = 0.72 \exp(-t/164), \; r^2 = 0.62 \tag{15}$$

with a time constant of 164 weeks. Their results over the first postnatal year do not conflict with the results previously discussed.

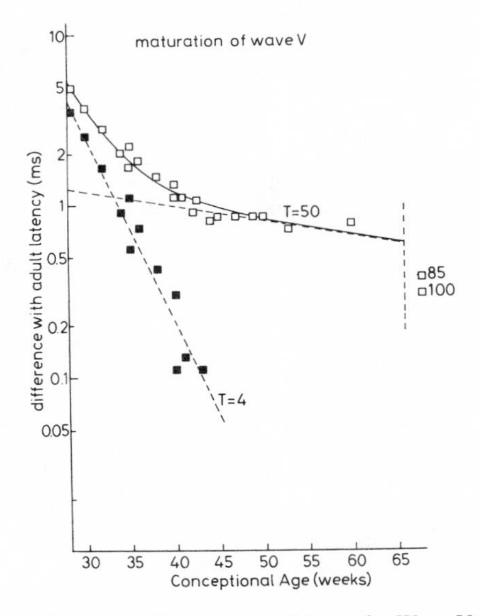

Figure 10. Fitting exponential curves to the changes in latency for Wave V (open symbols). A good fit is obtained by the sum of two exponential functions with time constants of 4 and 50 weeks. Solid symbols are the same data points for the early portion of the curve after correction for the contribution of the second (and longer) exponential component. The numbers beside the data points indicate that they have been obtained at 85 and 100 weeks, respectively.

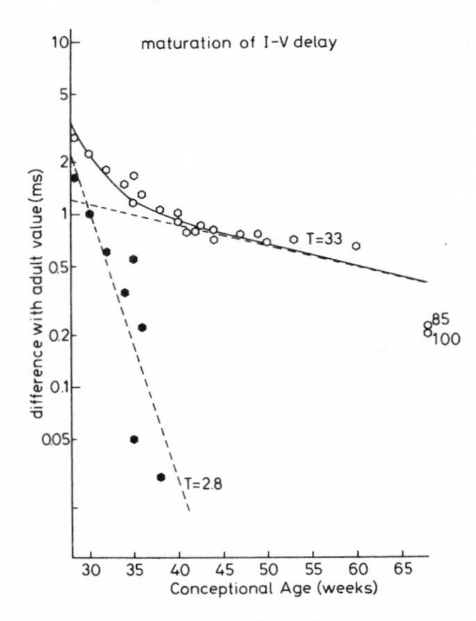

Figure 11. Changes in the I–V delay (open symbols) as a function of age. The data can be fitted with the sum of two exponential functions with time constants of 2.8 and 33 weeks. Solid symbols are the same data points for the early portion of the curve after correction for the contribution of the second (and longer) exponential component. The numbers beside the data points indicate that they have been obtained at 84 and 100 weeks, respectively.

The results of Hecox and Galambos (1974), extending over a 32-month period after normal birth, indicate an exponential decrease with a time constant of 32 weeks. Their Wave V latency changes, however, are much larger for early ages than for the data shown in Figures 8 and 10. Gafni, Sohmer, Gross, Weizman, and Robinson (1980) basically confirm the near completeness of maturation at normal birth for Wave I; their data for 42-weeks-conceptional age indicate a 0.3 ms difference from adult values. Recent normative data for preterm infants by Cox, Hack, and Metz (1981) are completely at variance with nearly all other data by indicating that the I–V latency difference in the preterm infant (34–40-weeks-conceptional age) is only 0.6–0.7 ms longer than in adults. The rate of change for their I–V difference over the first 4 months, however, would be characterized again by a time constant of 30–50 weeks.

Middle-Latency Potentials

The middle-latency components have not been studied extensively during the maturational process. One important reason might be that not much difference in latencies can be found as a function of age (Goldstein & McRandle, 1976). Mendelson and Salamy (1981) recently compared latency and amplitude values in 15 prematures (31–37 weeks), 15 fullterm infants (12 hours–4 days), 15 young children (3–4 years), and 15 adults (24–39 years). They found significant latency changes only for the P_0 component, which, incidentally, is Wave V from the BAEP. The actual middle-latency components showed only a shift in the peak-to-peak amplitudes, being maximal in young children.

Cortical Evoked Potentials

Cortical evoked potentials again show changes during development. Since these potentials are very dependent on the EEG as well as such factors as attention, restlessness, and habituation, findings are not always comparable. The latency of the various components is grossly influenced by the sleep stage of the infants, and Akiyama, Schulte, Schultz, and Parmelee (1969) could not show consistent maturational changes even within well-defined sleep stages. Ornitz, Ritvo, Lee, Panman, Walter, and Mason (1969) report that the latencies of Wave N_2 in babies (6–12 months) were similar to those for normal children during

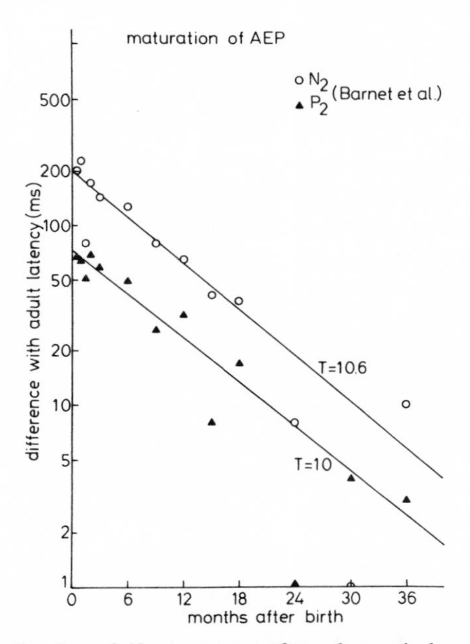

Figure 12. Changes in the P_2 and N_2 components from the cortical evoked potential follow a decreasing exponential (data from Barnet *et al.*, 1975). For the calculation of the N_2 curve, the data points (1.5 mo, 81) and (30 mo, 1) have been omitted.

sleep Stage 2. During REM sleep, however, the latencies were significantly longer in babies. Taguchi, Picton, Orpin, and Goodman (1969) found significant differences in N_2 latency but not for P_2 over three age groups: 0 to 2 days old, 2 to 4 days old, and over 4 days old.

Much more consistent data were obtained by Barnett, Ohlrich, Weiss, and Shanks (1975) in sleeping children from various age groups. They observed no significant latency changes for the early complex (N_0, P_1, N_1) but N_2 and P_2 latencies decreased with age. Replotting their data (Figure 12) reveals exponential changes for both components, with time constants of around 10 months (i.e., about 45 weeks). In a longitudinal study of 16 subjects, Ohlrich, Ritvo, Lee, Panman, Walter, and Mason (1978) confirmed these group data.

Discussion

From the quantitive information that has been obtained from the latency of the various waves from BAEP and cortical evoked potentials, one may attempt to summarize the time courses of maturation in the human cochlea, the brainstem, and the auditory cortex. Basically, there are three ranges of time constants obtained from the various curve fits:

1. Short time-constant maturation T_s of 3 to 5 weeks
2. Middle time-constant maturation T_m of 30 to 50 weeks
3. Long time-constant maturation T_l of about 164 weeks

With respect to the short time-constant processes, we must try to evaluate their appearance in the development of the I–V delay. From studies of high-pass noise masking (Don & Eggermont, 1978), it has become evident that the generators for Wave I are located more basally in the cochlea than the generators that ultimately give rise to Wave V. Thus, part of the I–V delay is actually a traveling-wave delay; this mechanical contribution is quite small for the normal human cochlea. In the maturing cat cochlea, Brugge *et al.* (1978) have demonstrated that the traveling-wave delay between, say, 1 kHz and 3 kHz is much greater than in the adult cochlea. Under the assumption that, as in the cat, the place–frequency relation is established at quite an early age and that the traveling-wave concept for the cat also holds for humans (Eggermont, 1976, 1979), we may also attribute the short time-constant process in the I–V delay to a cochlear mechanism. This would allow us to attribute all processes with time constants from 3 to 5 weeks to cochlear maturation.

An alternative hypothesis is that the short time-constant process in the I–V delay reflects the ultimate formation of the synaptic contacts, that is, by elimination of part of the innervation (Purves & Lichtman, 1980). This is a process that is virtually complete at normal birth (40-weeks-conceptional age).

The middle and long time-constant processes are both likely candidates to describe maturation of the central nervous system due to improved myelination. In contrast to the cat data, this central maturation in humans has a much longer time constant than the peripheral one.

General Discussion

We began our analysis of developmental changes in the auditory system of cats and humans with two simple assumptions: (1) all changes follow exponential curves; independently maturing parts of the system give rise to sums of exponential functions that describe the whole system, and (2) maturation in the peripheral part of the auditory system will be reflected in studies conducted at more central parts. For the cat, it appeared possible to describe most of the changes with a time constant of 6 or 12 days, depending on the set of latency data. This does not mean, however, that the changes in all parameters studied vanish at the same time. This is most evident from the latency changes for the various centers in the auditory system (Figure 4). Although there is hardly any difference in the time constants, the latency for the auditory nerve reaches adult values around 25 days after birth. For the cortical evoked potentials, about 37 days are needed.

Other models to describe latency changes, for example, have been used. Pujol (1972) used power curve fits of the form

$$L(t) = L(0)[t + A]^B$$

One nice result of such curve fits is the appearance of an asymptotic latency value $L(0)$, that is, the latency when the potentials are first detectable at $t = -A$. Pujol arrived at $t = -7$ days (i.e., 7 days before normal birth) for the cochlear nucleus, $t = -5$ days for the inferior colliculus, $t = -2$ days for the medial geniculate body and $t = 0$ days (birth) for cortical Area 1. This indicates that the system begins to function from the periphery. Our analysis adds that all the *changes* in each nucleus go equally fast, and are probably ruled by one and the same process: the peripheral input to the central nervous system.

In the cat, numerous changes take place in the PSTHs during the first month; this has been related to the immaturity of synapses. The ability to phase-lock the action potentials to tones also needs about a month to mature. Both phenomena might affect directional hearing, especially the part that is based on interaural time differences. As Moore and Irvine (1981) have also reported, the ability to make efficient use of interaural intensity differences takes some time to develop. We may extrapolate this to hearing in human neonates and expect that orientation reflexes to sound will take some time to become established. Clinicians using behavioral hearing tests with young children are well aware of this. From a comparison of the N_1 data in the cat and Wave I data in humans, one might conclude that the time constant for cochlear maturation in humans is about 2 to 5 times as large as in the cat (4 to 5 weeks compared to 6 to 12 days). Since it takes about a month for the cat cochlea to mature in nearly all aspects, one readily extrapolates this to, at most, 5 months in the human neonate. In the cat, however, the data are referenced with respect to birth, whereas the human data used in our analysis start at about 3 months before normal birth. As can be seen from Figure 4, the human cochlea, in general, will also reach maturity one month after normal birth. As a consequence, it should take at least one month after normal birth before directional hearing is mature and orientation responses are possible (not taking into account the required motor ability).

From an analysis of the available human BAEP data, it can be concluded that the model involving summed exponential functions cannot be rejected. It may allow us to separate cochlear maturation processes from central ones. It allows an estimate of the time course in humans for maturational processes that have only been studied so far in cats (e.g., directional hearing and frequency selectivity).

Since a number of changes in the parameters that characterize the functional status of the auditory system can be related to cochlear processes, we can draw parallels with studies in adult and pathological cochleas. Changes in threshold, spike rate, slope of the input–output curve, degree of tuning and latency of N_1 are all intimately related processes. Their changes with age are also partly reflected in cochleas in which reversible and irreversible hearing loss have been induced. A striking example was found in the relation between Q_{10dB} and threshold at the CF. This allows a particular interpretation of the cochlear maturation process as the increased effectiveness of the second filter, a concept that also plays a role in the understanding of phenomena associated with sensorineural hearing loss.

Besides the various effects that can be related to second-filter maturation, the more central changes are probably all related to changes in myelination of nerve fibers. By one month after normal birth, one does not expect synaptic changes or routing of nerve fibers to play a prominent role any more. The myelination process and the synaptic changes during the time that the cochlea matures are considered to proceed independently and their effect may be added; hence the summed exponentials.

The time constants for the maturation of the I–V delay in the brainstem auditory evoked potentials and the P_2 and N_2 components from the cortical potentials are the same, that is, about 30 to 50 weeks. Surprisingly, the middle components do not show any latency change; this calls into question their purported intermediate nature (Mendelson & Salamy, 1981) and rules out their usefulness in studying maturational processes.

The long time-constant (about 164 weeks) is based on only one survey and its existence as separate from the 30 to 50 week middle time-constant awaits further investigation. For practical purposes, the short and middle time-constant processes seem to be the most important ones.

References

Aitkin, L. M., & Moore, D. R. Inferior colliculus. II. Development of tuning characteristics and tonotopic organization in central nucleus of the neonatal cat. *Journal of Neurophysiology,* 1975, *38,* 1208–1216.

Akiyama, Y., Schulte, F. J., Schultz, M. A., & Parmelee, A. H. Acoustically evoked responses in premature and full term newborn infants. *Electroencephalography and Clinical Neurophysiology,* 1969, *26,* 371–380.

Barnet, A. B., Ohlrich, E. S., Weiss, I. P., & Shanks, B. Auditory evoked potentials during sleep in normal children from ten days to three years of age. *Electroencephalography and Clinical Neurophysiology,* 1975, *39,* 29–41.

Brugge, J. F., Javel, E., & Kitzes, L. M. Signs of functional maturation of peripheral auditory system in discharge patterns of neurons in anteroventral cochlear nucleus of kitten. *Journal of Neurophysiology,* 1978, *41,* 1557–1579.

Carlier, E., Abonnenc, M., & Pujol, R. Maturation des réponses unitaires à la stimulation tonale dans le nerf cochléaire du chaton. *Journal de Physiologie* (Paris), 1975, *70,* 129–138.

Carlier, E., Lenoir, M., & Pujol, R. Development of cochlear frequency selectivity tested by compound action potential tuning curves. *Hearing Research,* 1979, *1,* 197–201.

Carlier, E., & Pujol, R. Role of inner and outer haircells in coding sound intensity: An ontogenetic approach. *Brain Research,* 1978, *147,* 174–176.

Cox, C., Hack, M., & Metz, D. Brainstem-evoked response audiometry: Normative data from the preterm infant. *Audiology,* 1981, *20,* 53–64.

Dallos, P., & Cheatham, M. A. Compound action potential (AP) tuning curves. *Journal of the Acoustical Society of America,* 1976, *59,* 591–597.

Don, M., & Eggermont, J. J. Analysis of click-evoked brainstem potentials in man using high-pass noise masking. *Journal of the Acoustical Society of America,* 1978, *63,* 1084–1092.

Eggermont, J. J. Analysis of compound action potential responses to tonebursts in the human and guinea pig cochlea. *Journal of the Acoustical Society of America,* 1976, *60,* 1132–1139.

Eggermont, J. J. Compound action potential turning curves in normal and pathological human ears. *Journal of the Acoustical Society of America,* 1977, *62,* 1247–1251.

Eggermont, J. J. Narrow-band AP latencies in normal and recruiting human ears. *Journal of the Acoustical Society of America,* 1979, *65,* 463–470.

Eggermont, J. J., & Don, M. Analysis of click-evoked brainstem potentials in humans using high-pass noise masking. II. Effect of click intensity. *Journal of the Acoustical Society of America,* 1980, *68,* 1671–1675.

Evans, E. F. The sharpening of cochlear frequency selectivity in the normal and abnormal cochlea. *Audiology,* 1975, *14,* 419–442.

Fabiani, M., Sohmer, H., Tait, C., Gafni, M., & Kinarti, R. A functional measure of brain activity: Brainstem transmission time. *Electroencephalography and Clinical Neurophysiology,* 1979, *47,* 483–491.

Fernadez, C., & Hinojosa, R. Postnatal development of endocochlear potential and stria vascularis in the cat. *Acta Oto-laryngologica* (Stockholm), 1974, *78,* 173–186.

Gafni, M., Sohmer, H., Gross, S., Weizman, Z., & Robinson, M. Analysis of auditory nerve brainstem responses (ABR) in neonates and very young infants. *Archives of Otorhinolaryngology,* 1980, *229,* 167–174.

Goldstein, R., & McRandle, C. C. Middle components of the averaged electroencephalic response to clicks in neonates. In S. E. Hirsh, D. H. Eldridge, I. J. Hirsh, & S. R. Silverman (Eds.), *Hearing and Davis: Essays honoring Hallowell Davis.* St. Louis: Washington University Press, 1976.

Hecox, K., & Galambos, R. Brainstem auditory evoked responses in human infants and adults. *Archives of Otolaryngology,* 1974, *99,* 30–33.

Javel, E. Neurophysiological correlates of auditory maturation. *Annals of Otology, Rhinology and Laryngology,* 1980, *89* (Suppl. 74) 103–113.

Kiang, N. Y. S., Watanabe, T., Thomas, E. C., & Clark, L. F. *Discharge patterns of single fibers in the cats auditory nerve.* Cambridge: M.I.T. Press, 1965.

Mendelson, T., & Salamy, A. Maturational effects on the middle components of the averaged electroencephalic response. *Journal of Speech and Hearing Research,* 1981, *24,* 140–144.

Moore, D. R., & Irvine, D. R. F. The development of some peripheral and central auditory responses in the neonatal cat. *Brain Research,* 1979, *163,* 49–59.

Moore, D. R., & Irvine, D. R. F. Development of binaural input, response patterns, and discharge rate in single units of the cat inferior colliculus. *Experimental Brain Research,* 1980, *38,* 103–108.

Moore, D. R., & Irvine, D. R. F. Development of responses to acoustic interaural intensity differences in the cat inferior colliculus. *Experimental Brain Research,* 1981, *41,* 301–309.

Ohlrich, E. S., Barnet, A. B., Weiss, I. P., & Shanks, B. L. Auditory evoked potential development in early childhood: A longitudinal study. *Electroencephalography and Clinical Neurophysiology,* 1978, *44,* 441–423.

Ornitz, E. M., Ritvo, E. R., Lee, Y. H., Panman, L. M., Walter, R. D., & Mason, A. The auditory evoked response in babies during REM sleep. *Electroencephalography and Clinical Neurophysiology,* 1969, *27,* 195–198.

Pujol, R. Développement des réponses à la stimulation sonore dans le colliculus inférieur chez le chat. *Journal de Physiologie* (Paris), 1969, *61,* 411–421.

Pujol, R. Development of toneburst responses along the auditory pathway in the cat. *Acta Oto-laryngologica* (Stockholm), 1972, *74,* 383–391.

Purves, D., & Lichtman, J. W. Elimination of synapses in the developing nervous system. *Science,* 1980, *210,* 153–157.

Riggs, D. S. *The mathematical approach to physiological problems.* Cambrige: M.I.T. Press, 1970.

Robertson, D., & Manley, G. A. Manipulation of frequency analysis in the cochlear ganglion of the guinea pig. *Journal of Comparative Physiology,* 1974, *91,* 363–375.

Romand, R. Maturation des potentiels cochléaires dans la période périnatale chez le chat et chez le cobaye. *Journal de Physiologie* (Paris), 1971, *63,* 763–782.

Romand, R. Development of auditory nerve activity in kittens. *Brain Research,* 1979, *173,* 554–556.

Romand, R., Granier, M. R., & Marty, R. Développement postnatal de l'activité provoquée dans l'olive supérieure latérale chez le chat par la stimulation sonore. *Journal de Physiologie* (Paris), 1973, *66,* 303–315.

Romand, R., & Marty, R. Postnatal maturation of the cochlear nuclei in the cat: A neurophysiological study. *Brain Research,* 1975, *83,* 225–233.

Rubel, E. W. Ontogeny of structure and function in the vertebrate auditory system. In M. Jacobson (Ed.), *Handbook of sensory physiology* (Vol. 9). New York: Springer Verlag, 1978.

Ruben, R. J., & Rapin, I. Plasticity of the developing auditory system. *Annals of Otology, Rhinology and Laryngology,* 1980, *89,* 303–311.

Salamy, A., & McKean, C. M. Postnatal development of human brainstem potentials during the first year of life. *Electroencephalography and Clinical Neurophysiology,* 1976, *40,* 418–426.

Saunders, J. C., Dolgin, K. G., & Lowry, L. D. The maturation of frequency selectivity in C57BL/6J mice studied with auditory evoked response tuning curves. *Brain Research,* 1980, *187,* 69–79.

Shah, S. N., & Salamy, A. Auditory-evoked far-field potentials in myelin deficient mutant quaking mice. *Neuroscience,* 1980, *5,* 2321–2323.

Shah, S. N., Bhargava, V. K., & McKean, C. M. Maturational changes in early auditory evoked potentials and myelination of the inferior colliculus in rats. *Neuroscience,* 1978, *3,* 561–563.

Shipley, C., Buckwald, J. S., Norman, R., & Guthrie, D. Brainstem auditory evoked response development in the kitten. *Brain Research,* 1980, *182,* 313–326.

Shnerson, A., & Willott, J. F. Development of inferior colliculus response properties in C57BL/6J mouse pups. *Experimental Brain Research,* 1979, *37,* 373–385.

Starr, A., Amlie, R. N., Martin, W. H., & Sanders, S. Development of auditory function in newborn infants revealed by auditory brainstem potentials. *Pediatrics,* 1977, *60,* 831–839.

Taguchi, K., Picton, T. W., Orpin, J. A., & Goodman, W. S. Evoked response audiometry in newborn infants. *Acta Oto-laryngologica. Supplementum* (Stockholm), 1969, *252,* 5–17.

Uziel, A., Marot, M., & Germain, M. Les potentiels évoqués du nerf auditif et du tronc cérébral chez le nouveau-né et l'enfant. *Revue de Laryngologie* (Bordeaux), 1980, *101,* 55–71.

Walsh, E. J., McGee, J., & Javel, E. Development of BSER in kittens: Evidence for different central and peripheral maturation rates. *Journal of the Acoustical Society of America,* 1981, *69,* S84, (Suppl. 1).

Willot, J. F., & Shnerson, A. Rapid development of tuning characteristics of inferior colliculus neurons of mouse pups. *Brain Research,* 1978, *148,* 230–233.

Yost, W. A., & Nielson, D. W. *Fundamentals of hearing.* New York: Holt, Rinehart & Winston, 1977.

COMMENTARY

Introductory Comments on the Anatomy and Physiology of the Developing Auditory System

Ivan Hunter-Duvar

E. N. T. Department
The Hospital for Sick Children
555 University Avenue
Toronto, Ontario

A symposium that concerns itself with the development and testing of hearing in infants requires accurate information about the anatomy of the hearing apparatus. At the place and point in time when human fetal material was available, Dr. Bredberg performed a classic study on the development of the inner ear. Changes in medical and legal procedures make it unlikely that this fine piece of work will ever be duplicated.

Dr. Bredberg has combined his findings with previous results of Bast and Anson to demonstrate that the human inner ear is anatomically mature before birth. Further evidence is presented to indicate that the hearing apparatus is functional before birth. Information on the degree of function at birth and thereafter is critical to those wishing to measure the efficacy of the auditory system.

Dr. Eggermont, in brilliant fashion, has put together a great deal of the animal electrophysiological data relevant to the auditory system and has shown that the results fit an exponential development model. Much of the data come from the cat. The comparison with human data must take into consideration that anatomically the cochlea in the former is not mature until four weeks after birth whereas the latter is anatomically mature approximately 12 weeks before birth.

Electrophysiological data for human neonates are largely confined to brainstem auditory evoked potentials. Dr. Eggermont demonstrates that these data can be handled with a two-exponential fit and suggests, for humans, the possibility of a short time-constant for cochlear maturation and a longer time-constant to reflect central maturation. A comparison of human data to the animal model leads to a suggestion that it will be one month after normal birth before orientation responses are possible in the human neonate. This hypothesis undoubtedly will be reviewed with interest by behavioral psychologists and audiologists involved in investigations of infant hearing.

PART II

BASIC AUDITORY PROCESSES IN INFANCY

CHAPTER 3

The Development of Infants' Auditory Spatial Sensitivity

Darwin W. Muir

Department of Psychology
Queen's University
Kingston, Ontario

Introduction

Early research in the area of perceptual development has been inspired by two opposing viewpoints: nativism and empiricism. Extreme nativists argue that a perceptually naive infant possesses the perceptual abilities of experienced adults, whereas empiricists such as William James maintain that the young infant's world is composed of a "blooming, buzzing confusion." According to this latter view, objects and people gradually emerge through experience from the background of perceptual chaos.

Today, no theorist holds such extreme views; nevertheless, most modern theories can be seen as biased in favor of one of these two positions. An example of a recently formulated nativist theoretical position is that of Bower (1974, 1979), who believes that a primitive unity of perceptual space exists at birth. He suggests that newborns are capable of responding to certain abstract, amodal properties of stimuli, such as their intensity, movement, tempo, and general location in space, but they are unable to tell whether the input consists of lights,

This research was supported by a grant to Darwin W. Muir and Peter C. Dodwell from the National Science and Engineering Research Council and by a Social and Behavioral Research Grant to Darwin Muir from the National Foundation—March of Dimes.

sounds, smells, or touches. This ability to distinguish and respond differentially to input from the different sense modalities is what emerges in development. The idea that behavior differentiates from the general to the specific has a long history in embryology (e.g., Coghill, 1929) and developmental psychology (Werner, 1934).

As support for his modern version of differentiation theory, Bower points out that newborns will move their eyes in the direction of off-centered sounds (e.g., Crassini & Broerse, 1980; Mendelson & Haith, 1976; Turkewitz, Birch, Moreau, Levy, & Cornwell, 1966; Wertheimer, 1961). Thus, he argues that some degree of intersensory coordination is present at birth. He also contends that because newborns and very young infants cannot tell the difference between sights and sounds, a true intersensory equivalence or substitution is possible at an early age prior to sensory differentiation. To support this idea, Bower (1977) presented evidence that a 4-month-old, congenitally blind baby could "see with his ears." His subject wore an ultrasonic echo-location device that presented him with a stereophonic image of silent objects at different positions in space. Bower reports that the baby tracked and reached for moving objects as soon as the aid was turned on.

By contrast, Piaget (1952), expounding a more generally accepted viewpoint, emphasizes the role of early experience in the ontogeny of space perception. According to his theory, little perceptual organization exists at birth. The few reflexes that can be elicited by auditory, visual, and tactile stimulation exist in an unrelated and unorganized fashion. Piaget observed that his babies crudely localized visual targets, but not sounds, during their first month of life. They began to orient toward sounds in their second month. And between their third and sixth month, through the process of "reciprocal assimilation," they began to develop a sense of a multimodal, unitary space as they actively experienced natural correlations between sights, sounds, and touches (e.g., manipulating a noisy rattle or looking for a face to match a voice). This has become known as the integration model of perceptual development.

Piaget's observations agree with those of audiologists and psychologists who, over the last 20 years, have documented the average age at which various auditory localization responses are first observed. The results of five representative and comparable studies are summarized in Table 1. The children sat on their mothers' laps facing an experimenter who judged whether or not the babies turned toward a noise coming from the left, right, above, or below their line of sight. Clearly, their results agree that infants begin to orient toward sounds along the horizontal plane between 3 and 6 months of age and to sounds in the vertical median plane after 6 months of age. For example, Watrous, McConnell, Sitton, and Fleet (1975, Table 4) report that responses in the horizontal plane peaked at 6 to 8 months of age, whereas vertical localization responses peaked at 9 to 12 months of age. Northern and Downs (1978) go on to state that infants do not

Table 1. Ages at Which Infants Were First Observed to Localize Sounds Along the Horizontal and Vertical Median Plane

Authors	Localization responses	
	Horizontal plane	Vertical plane
Bayley (1969)	4–5 months	
Uzgiris and Hunt (1975)	3–5 months	
Chun, Pawsat, and Forster (1960)	5–6 months	7–8 months
Watrous, McConnell, Sitton, and Fleet (1974)[a]	3–5 months (14%)	0%
	6–8 months (49%)	8%
	9–12 months (29%)	32%
Northern and Downs (1978)	4–7 months	9–13 months

[a]Watrous *et al.* (1974) give the percentages of response elicited at different ages.

suddenly begin to orient precisely toward sounds above and below the horizontal plane: first, they localize sounds below it (between 9–13 months), then, both above and below it (between 16–21 months); and finally, after 21 months, they directly locate sound at any angle.

These planar differences are readily accounted for within a Piagetian context simply because the vertical and horizontal coordinates of a sound's position in space are determined by different cues. Searle, Braida, Davis, and Colburn (1976) described six cues in their model of adult auditory localization. The major cues for determining the apparent azimuth are binaural. They include the direction-dependent change as a sound moves around the head in interaural time delay (caused by sound waves arriving at one ear before the other) and interaural intensity difference (due to head shadow). Monaural head shadow cues also relate to azimuthal position. A different set of cues is used to determine the apparent elevation of sounds on the vertical median plane where no interaural time or intensity differences exist. For sounds in this plane, the external ear generates two independent cues, differential frequency responses and binaural pinna disparities, which change systematically with a sound's angle of elevation. Given the complexity of the cues used to represent a sound image in auditory space, it is not surprising to find that infants begin to locate a sound's vertical and horizontal coordinates at different ages.

Extending the above argument, one would not expect very young infants to experience spatial equivalence, as Bower postulates. On the other hand, they do appear to make some spatial responses to stimuli from several modalities, including audition (e.g., Dayton & Jones, 1964; Rieser, Yonas, & Wikner, 1976; Turkewitz, Gordon, & Birch, 1965; Turkewitz *et al.*, 1966). The literature on early intermodal spatial coordination prompted the research to be described in this chapter, but it will not be considered in detail here. (For comprehensive reviews of intermodal coordination see Butterworth, 1981; Pick, Yonas, & Rieser, 1979.)

The focus of this chapter is on the development of one spatial response: the infant's natural tendency to turn and face sounds at different locations in space. The presence of such a response implies the existence of some form of spatial representation. Thus, Piaget's and Bower's ideas can be evaluated against this background. The chapter is subdivided into three sections. The first considers localization responses of newborns and young infants to sounds on the horizontal plane; the second looks at localization to sounds on the vertical plane; and the third describes the development of localization responses to sounds by infants with hearing losses and by a totally blind infant who wore a sonar aid.

Responses to Sounds in the Horizontal Plane

A number of investigators have reported that newborns, with their heads held firmly, reflexively flick their eyes toward a brief click presented next to one of their ears (Turkewitz *et al.*, 1966; Wertheimer, 1961). Wolff (1959) and Hammond (1970) also presented anecdotal evidence that newborns would turn toward sounds. It should be noted that experimental work on the newborn's ability to actually turn and face a sound source is of recent origin. Brazelton (1973), convinced that newborns localize sounds, included auditory localization items in his clinical newborn examination. The baby is held by the examiner who shakes a rattle next to each ear or calls to the baby from the side and records whether or not the baby responds by blinking, brightening, shifting eyes and head toward the sounds, and visually scanning the stimulus. Brazelton and his co-workers report that many infants will turn and look at sound sources; however, a number of uncontrolled factors such as visual cues and examiner bias may have contributed to their results.

General Methodology and Newborn Localization Abilities

We have tested newborns under more controlled laboratory conditions in order to demonstrate unequivocally that they would reliably locate sounds. Our auditory stimulus is a rattle sound produced by rhythmically shaking a hospital specimen bottle half-filled with popcorn at the rate of about 2 shakes per sec. In some studies, we use the real rattle, and in others, recorded rattle sounds are presented through stationary speakers. Spectrum analysis indicates that the recorded rattle sound is a broad-band stimulus having a reasonably flat energy distribution between 10 and 13,000 Hz with a peak at 2,900 Hz. The average intensity of the sound used in most studies is approximately 80 dB (SPL) against a background level of about 60 dB.

In a test session, we examine the babies while they are in a quiet, alert state, usually 1 to 2 hr following a feeding. The general procedure consists of one experimenter holding the baby in a semisupine position facing the ceiling so that the head and back are supported but unrestrained head movements are permitted. A trial begins when the baby's head is in the midline; the rattle sound is presented approximately 20 cm from one of the baby's ears for 20 sec or until the baby completes a head turn to one side or the other. When the real rattle is used, an empty container is shaken opposite the other ear to balance tactile and visual cues. Between trials, the baby is gently rocked to maintain an alert state and then repositioned for the next trial. To eliminate experimenter bias, in most studies the person holding the infant wears headphones through which the rattle sound is presented simultaneously to both ears and thus is unaware of the location of the sound because the stereo sound image appears to be located in the middle of the holder's head. The method of handling is illustrated in the two photographs shown in Figure 1. Eight to 16 trials are given, depending on the experiment, and the baby's performance is videotaped for later analysis by observers who do not know the location of the sound. We record the direction of the first head movement and that of the largest head turn (usually they are the same) as well as the response latencies.

The results of several of our studies suggest that newborns are slow to respond, taking an average of 2 to 3 sec to initiate a head turn and about 5 sec to complete it. However, they generally are remarkably accurate, turning toward the sound source in most experiments on an average of 75% to 90% of the trials. The performance of individual babies in our first study (Muir & Field, 1979), shown in Table 2, is typical. Here, 11 out of 12 subjects turned more often toward the sound than away from it, and very few subjects failed to turn on stimulus trials. By contrast, during silent control trials, the babies failed on 40% of the trials, and when they did turn, it was usually toward their right side. Also, in this experiment, the extent of head rotation was classified as falling into one of 4 equal segments (22½°) within 90° on either side of midline. As shown in the table, most babies turned completely around to face the sound, which was much further than they did on silent control trials.

Babies will orient to sound within minutes after delivery (Muir, Abraham, Forbes, & Harris, 1979). In fact, in our experience, newborns appear more alert and responsive during their first postnatal hours than they do for the next few days. However, most of our subjects have been 3 to 4 days old when tested. We have also been conducting a screening program for the early identification of handicaps among preterm infants with birthweights < 1500 g (Muir, 1982). Forty-three infants were healthy enough to be tested during their preterm period

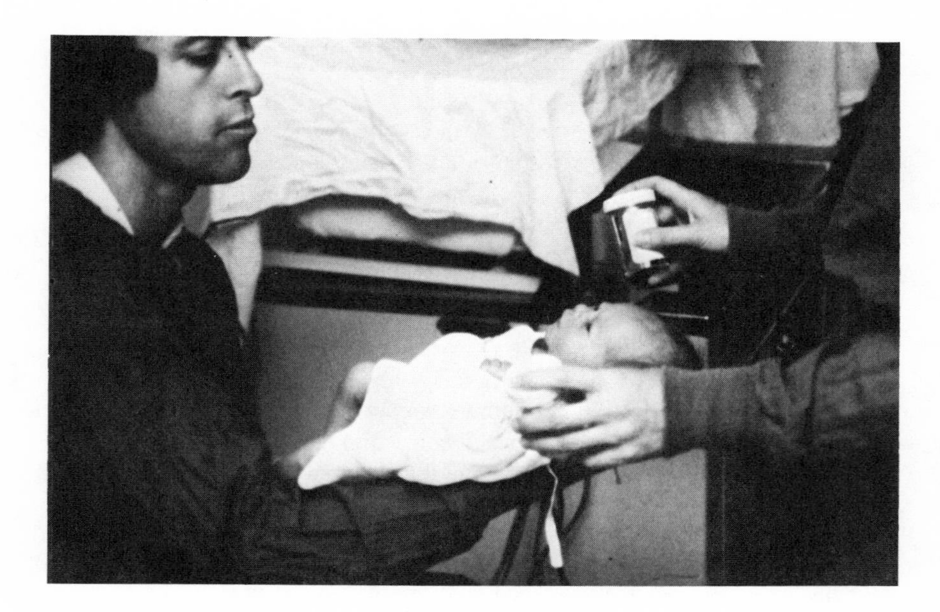

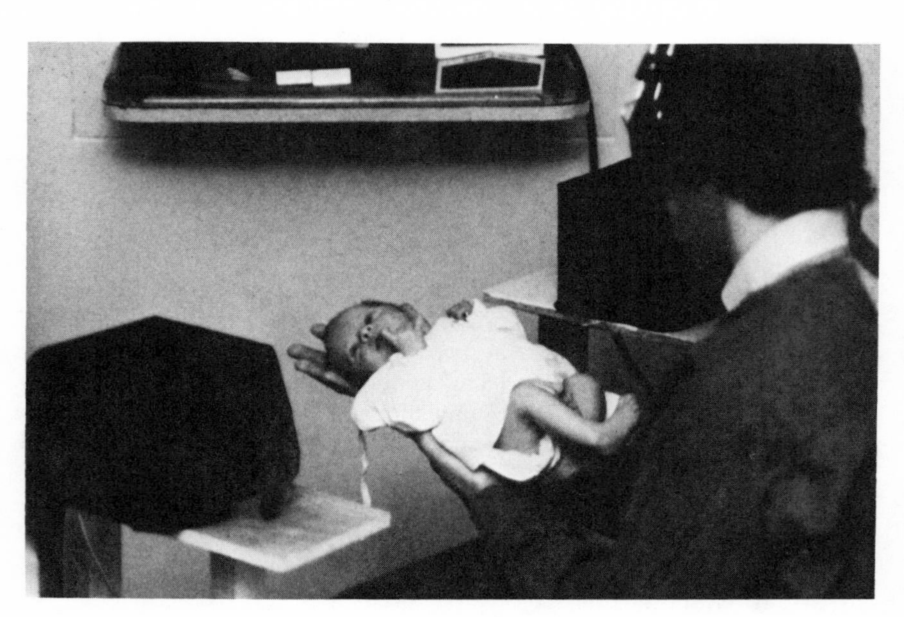

Figure 1. (A) An infant is held while experimenter shakes a noisy rattle on one side and a silent container on the other. (B) Sound is presented through the loud speaker on the infant's right side.

Table 2. Results of Head Turning in Experiment 2

	Experimental trials				Control trials			
	Number of maximum turns				Number of maximum turns			
Subject	Toward sound	Away from sound	No turn	Mean extent[a]	Left	Right	No turn	Mean extent
1	7	1	0	3.6	0	2	1	3.5
2	7	1	0	3.9	0	3	0	3.7
3	8	0	0	3.7	0	2	1	2.0
4	7	1	0	3.6	1	0	2	1.5
5	7	1	0	4.0	0	2	1	1.0
6	5	2	0[b]	3.5	1	1	0	1.8
7	7	1	0	4.0	1	1	1	3.0
8	3	3	2	1.0	1	2	0	1.7
9	6	2	0	3.7	0	2	1	3.5
10	5	1	2	1.2	1	0	2	1.0
11	3	2	3	2.7	1	0	2	2.0
12	7	1	2[c]	2.8	0	0	3	0.0

[a]Each maximum head turn was classified on a four-point scale (see text).
[b]One experimental and one control trial were lost due to a videotaping error.
[c]Two extra trials were given because the subject was unresponsive twice but remained in a good state.

when their average gestational age (GA) was estimated to be 35.4 weeks (range 32–36 wks GA), and they performed almost as well at that age as they did when they were retested on their expected birthdate (average GA= 40.3 weeks, range 39–43 weeks). In this study, 8 auditory localization trials were given using real rattles and the procedure described above, except that if no turn occurred the trial lasted for 30 sec, rather than 20 sec. The median percentage of trials on which infants turned their heads toward the sound was 75% on the preterm test and 88% on the fullterm test. There were two obvious differences between preterm and fullterm performance. On the preterm test, these infants took longer to complete a response (median latency of 12 sec compared with 7 sec at term) and failed to turn on a larger number of trials (16% as compared with 2% at term).

Newborns also show some precision in pointing their heads at a sound source. Forbes, Abraham, and Muir (1979) tested 3-day-olds using the standard procedure, except that on each trial the recorded rattle sound was presented through 1 of 5 speakers arranged on a perimeter. One speaker was directly in front of the baby and the others were at 45° and 90° on either side of the baby's midline. The babies' head movements were recorded by videotaping the crowns of their heads. A thin black line was placed along the midsagittal suture so that the degree of head rotation from the midline could be estimated using a protractor placed on the videoscreen. We found that the maximum degree of head rotation in the direction of the sound source averaged 44° toward the 45° speakers and 60° toward the 90° speakers.

In this study, testing was carried out in almost complete darkness with low-light recording equipment and infrared illumination to eliminate visual interference (see Muir, 1982, for further details). Visual interference can easily occur as Fisher-Fay (1981) demonstrated when she used Forbes *et al.*'s (1979) testing and scoring procedure and presented a visual target (a circular pattern of light-emitting diodes) 30° to one side of midline at the same time as the rattle sound was presented 90° on the other side of midline. Under these conditions, only 36% of the first head turns were directed toward the sound source, whereas 78% and 77% of the first head turns were toward the auditory and visual stimuli, respectively, when they were presented alone. Consequently, in our current studies, we use very low-light levels.

The head orientation response toward laterally presented sounds is one of the most dramatic and reliable behaviors (next to the classical reflexes) that one can elicit from newborns provided proper testing conditions are used. One must eliminate competing visual stimuli, hold the baby properly, and present sounds long enough for infants to complete their response. For example, in studies where babies were placed in inclined infant seats, they turned toward sounds on fewer trials (e.g., 64% correct in Table 3 of Alegria & Noirot, 1978) or turned less far (Turner & Macfarlane, 1978) compared to studies in which they were held by an experimenter (e.g., 95% correct in Clifton, Morrongiello, Kulig, & Dowd, 1981). Other investigators employ stimuli that may be too brief. For example, the auditory trials in the Einstein Neonatal Neurobehavioral Assessment Scale (Kurtzberg, Vaughan, Daum, Grellong, Albin, & Rotkin, 1979) last for only 5 sec. Clearly, some of our full-term infants and most of our preterm infants who turned toward sounds on almost every trial would have failed Kurtzberg *et al.*'s test simply because they were too slow, given that they can take as long as 25 to 30 sec to respond in our tests. Finally, in our experience, performance is better if one avoids inserting too many control trials (either central sound or silent trials) and if the trial ends once the baby completes a head turn. Otherwise, infants tend to fuss after a few trials.

Responses to Sound in the Horizontal Plane by Older Infants

The fact that many newborn infants reliably orient toward sounds appears to contradict the extensive clinical and normative literature that suggests babies do not localize sounds on the horizontal plane until 5 or 6 months of age. To ascertain whether these other investigators simply failed to use a test procedure appropriate for younger infants, we conducted several longitudinal studies using our newborn testing procedure (Field, Muir, Pilon, Sinclair, & Dodwell, 1980; Muir *et al.*, 1979). A typical set of results, from Muir *et al.* (1979), is shown in Figure 2. The percentage of maximum head turns directed toward the sound is plotted as a function of age, in 20-day blocks, for each of 4 infants who were assessed biweekly from birth until they were 4 months old. Three of the infants (A, B, and D) turned to sound reliably at birth and during their first 20 to 40 days of life, performed poorly throughout their second and third months, and almost perfectly again when they were 4 months old. The fourth infant (C) never turned reliably until she was 4 months old. In this and other studies, we have found that some babies turn in the wrong direction at 2 months of age, whereas others simply stop turning to the sound. Throughout this period of decline in the localization response, the median latency for infants who do complete a head turn is approximately 7 sec. Then, when the babies begin to turn accurately to face sounds at around 4 months of age, their latency to complete a response is less than 2 sec.

Why infants should lose their newborn competence remains a mystery. We have tested several potential hypotheses. In case the infants were bored with the rattle, attempts were made to reinstate responding by presenting both novel (e.g., rattling paper) and meaningful sounds (e.g., mothers' voices), all to no avail (Muir *et al.*, 1979). The 2-month-olds would smile and vocalize when sounds

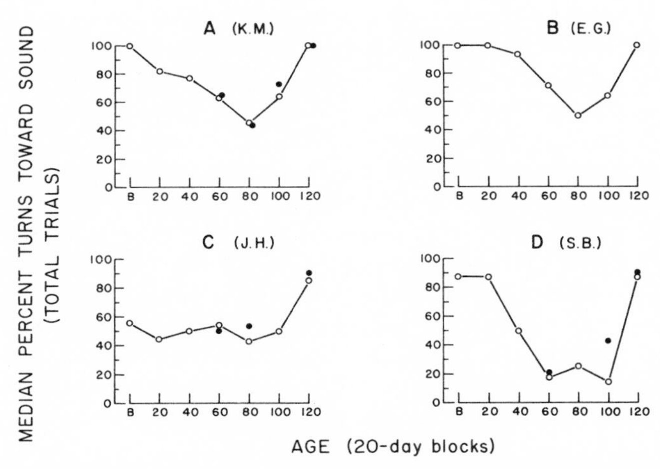

Figure 2. Median percentage of trials on which 4 infants turned toward the rattle sound, as a function of their age, collapsed into 20-day blocks.

were presented but failed to turn toward them (Muir, 1982). Also, many investigators (e.g., Haith, 1978) have noted that infants become more visually active at about 2 months of age and tend to become "captured" by visual stimuli. Therefore, we tested 3 of the 4 infants, whose data are shown in Figure 2 (A, C, and D), in complete darkness using infrared sensitive video recording equipment. Their performances under these conditions (represented in the figure by solid circles) were not influenced by lighting conditions, either during or following the period of poor responding.

The fact that newborns are capable of turning toward sounds is in agreement with the literature on other neonatal perceptual abilities. Crude localization occurs to off-centered visual (e.g., Dayton & Jones, 1964) and tactile (e.g., Turkewitz *et al.*, 1965) stimuli. Also, the structures that respond to interaural time and intensity differences at the brainstem and midbrain level of the major auditory pathway are present and apparently functional at birth (Hecox, 1975).

Although a number of explanations can be offered to account for the temporary decline in sound localization responses by infants during their second and third months of life (see Muir *et al.*, 1979), its existence suggests that the newborn response may belong to the same class of behaviors as a number of neonatal reflexes. The Moro, stepping, and palmar grasp reflexes diminish during the early months of life (e.g., McGraw, 1943). The loss of these reflexes is commonly presumed to reflect the maturation of cortical function which, as it gains control, begins to inhibit and modulate subcortical centers (e.g., Drillien & Drummond, 1977). Viewing the decline in auditory localization responses as a maturational phenomenon agrees with preliminary results from our high-risk follow-up study of 45 preterm infants who were born at least 6 weeks prior to their expected delivery date. If the decline and subsequent recovery were simply a product of extrauterine experience, we might expect them to appear at an earlier conceptional age in this group. However, this was not the case. When the premature infants were tested at 40-weeks-conceptional age, and again at 3- and 6-months postterm, the percentages of head turns they directed toward the sound at the three ages were 76%, 40%, and 91%, respectively. Seventy-seven percent of the infants performed better on their due date than they did three months later. Thus, their performance decrement was timed to their age from conception, rather than their chronological age, which supports a maturational explanation.

Finally, it is important to note that the decline in localization performance so consistently observed in 2- to 3-month-old babies does not signify that at this age they are not able to detect and respond to the spatial location of sound sources. Indeed, under our test conditions, when infants of this age do turn their heads in the presence of a lateral sound, they turn toward it on the majority of trials, approximately 70% (Field *et al.*, 1980; Field, DiFranco, Dodwell, & Muir, 1979; and Muir *et al.*, 1979).

Responses to Sounds in the Vertical Plane

Adults use different sets of cues to locate sounds on the horizontal (interaural time and intensity differences) and vertical (pinna cues) planes. One might argue that although neonates relate the lateral positions of sounds to binaural disparities from birth, they may need experience to associate a sound's spectral properties with its elevation. On the other hand, just as was the case with sound localization in the horizontal plane, it is possible that a careful analysis might reveal that infants will orient toward sounds off the horizontal plane much earlier than 8 to 9 months of age as Northern and Downs (1978), Watrous *et al.* (1975), and others have suggested. We chose to investigate the vertical localization abilities of 4½-month-olds because this was the earliest age, except for birth, at which accurate horizontal localization responses could be readily elicited.

Vertical Localization Responses of 4½-Month-Olds

In our first experiment (Muir, unpublished) on localization in the vertical plane, we used a simple variant of our standard procedure. The baby was seated comfortably on the mother's lap facing a low-light camera placed above a mechanical toy. Trials were presented in darkness to avoid potential visual interference. Prior to each trial, the toy was illuminated and activated, and while the baby was centered, looking at it, the experimenter surreptitiously moved a rattle and an identical (but silent) container into position. When the toy was turned off, the two containers were rhythmically shaken for 10 sec or until the baby completed a head turn of at least 45°. Two blocks of five trials were given and, within a block, the rattle sound was presented from each of the following five locations: approximately 60° above and below the line of sight on the vertical median plane, approximately 60° left and right of the line of sight on the horizontal plane, and directly behind the baby's head. The control rattle was shaken in the position opposite the sound (e.g., with "above" sounds the empty container was below) during a trial to balance nonauditory stimuli.

Thirteen 4½-month-old infants were tested, and the data for the 10 who successfully completed a session were analyzed. The infants' first major head turn ($> 10°$) was classified by naive observers, as falling into one of the four 90° sectors surrounding the vertical and horizontal planes (up, down, left, or right). Interobserver agreement on this measure was 95% and on the other measures discussed below at least 89%. In Figure 3A, the percentage of trials on which infants rotated their heads in the different directions is given as a function of the

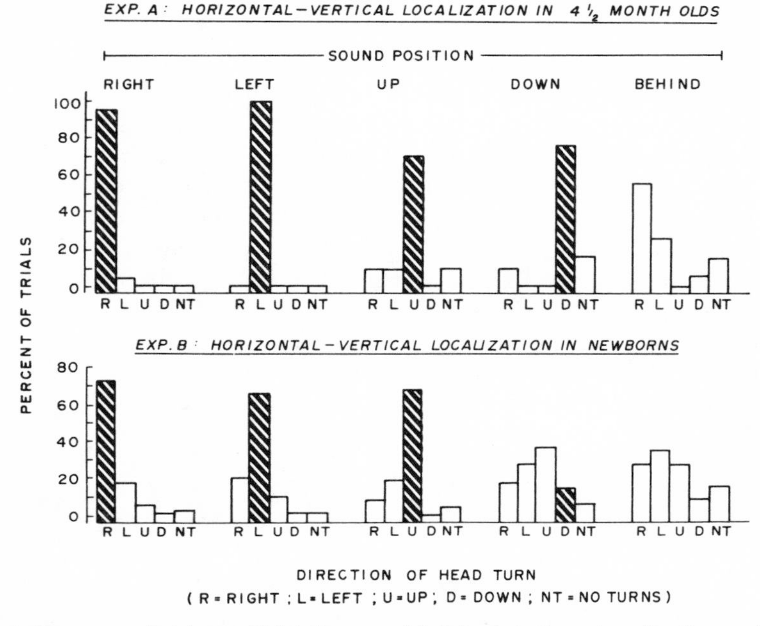

Figure 3. Percentage of trials on which babies turned their heads in the various directions or did not turn when the rattle was sounded from different directions relative to their line of sight.

position of the sound source (indicated by the solid bars). Clearly, these babies oriented toward the sound on almost every trial when it was left or right and on approximately 3 of the 4 trials when it was above or below the horizon. Eight of the 10 babies oriented on the majority of trials toward the sound when it was on the vertical plane and the other two were correct on 2 of the 4 trials. This is significantly better than chance, using 25% as an estimate. If trials where sound is behind the infant are used to estimate chance, it is clear that they do not distribute their responses equally, but usually turn to the right and almost never up or down. Thus, either estimate leads to the conclusion that 4½-month-olds do orient to sounds on the vertical plane. The condition with the sound behind the head was included as a midline control instead of a position dead ahead because babies tended to interrupt trials by kicking and reaching for sounds directly in front. It turns out that they also recognize that the sound is behind them and turn around much farther on these trials than on the lateral ones and at times reach out and grasp the rattle.

Although the babies turned reliably toward sounds along the vertical median plane, responses were less accurate than to sounds on the horizontal plane. Also 9 out of 10 infants responded more slowly in the vertical plane (average of 4.3

sec) than in the horizontal plane (average of 1.65 sec). They were slower orienting toward the sound behind their heads (average of 4.0 sec) as well. These longer latencies possibly represent less certainty about the target's location. This idea of initial confusion seems plausible when one considers the number of head oscillations the babies made prior to completion of their major head turns. These are brief, small head movements away from the center position (> 5°–10°) and back to the central position without a pause. This behavior was infrequent when sound was left or right (average of 1.2 oscillations/trial) but was common when it was up, down, or behind (average of 2.5 oscillations/trial).

Quantitative Measures of Vertical/Horizontal Head Rotations

Forbes (Forbes, 1981; Forbes & Muir, 1981) used a more elaborate procedure to quantify head rotation to sounds in oblique, as well as horizontal and vertical median, planes. He studied 15 babies between 4 and 4½ months of age, 10 of whom turned reliably to laterally presented sounds and were able to complete two successive test sessions. Their responses to sounds along different planes were assessed using the apparatus illustrated in Figure 4. Four 11.5 cm speakers (type CTS SR.10), enclosed in sealed plywood cabinets, were attached to a circular steel hoop at positions ±30° and ±60° from the center of rotation. The hoop was supported directly in front of and behind the infant's line of sight so that it could be rotated away from the horizontal plane (H) ±45° into the left and right oblique planes (OL and OR) and 90° into the vertical median plane (V).

During testing the mother sat on a chair in the center of the hoop holding her baby, whose head was 1.5 m from each speaker. The mother was made unaware of the stimulus location by wearing earphones and earplugs. Each test session consisted of 20 trials. On a sound trial, the recording of the rhythmically shaking rattle was played through 1 of the 4 speakers at 72 dB SPL (against a background of 58 dB) for 10 sec or until the baby's head turned and held the new position for at least 2 sec. The sound was presented once at each speaker location with the hoop in each of the 4 planes: H, V, OL, and OR. One 10-sec silent control trial was also given in each plane. The order of presentation of the planes, as well as the location of stimuli within a plane, was random. Prior to each trial, a centering stimulus, a blinking toy, oriented the baby's head and eyes directly toward an infrared sensitive videocamera. This frontal view of the baby's head-turning activity was monitored and videotaped for later analysis by an experimenter in an adjacent room, who presented the stimuli. Each infant was given 2, 20-trial test sessions on separate days. In one session, a standard Christmas tree light behind the baby provided illumination throughout testing and in the other session this light was on between trials and turned off during a trial. The order of lighting conditions was random. This allowed us to compare the effect of providing infants with a visible frame of reference.

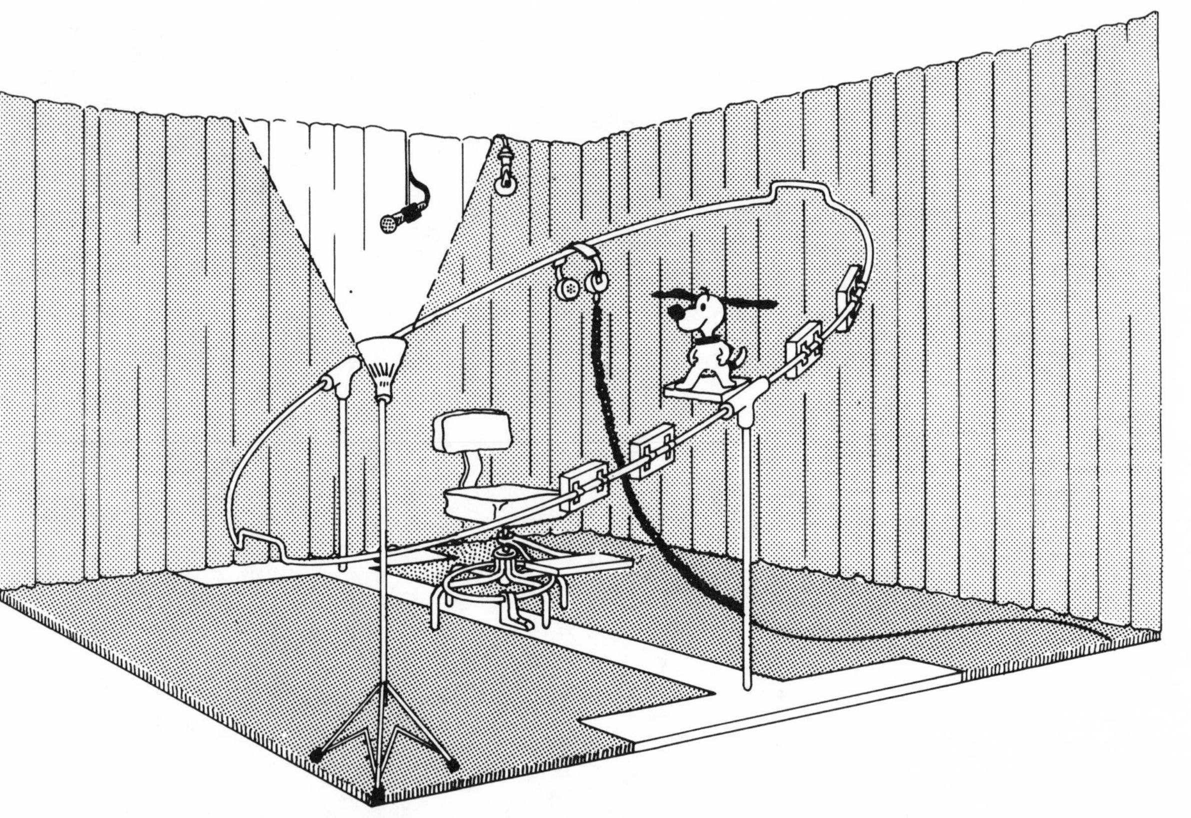

Figure 4. Drawing of the apparatus used by Forbes (1981) and Muir and Harris (1980) to test the localization responses of infants to sounds presented in vertical, horizontal, and oblique planes from speakers mounted on a rotatable steel hoop. The size of the centering toy, at the pivot point of the hoop, was 1½ in. and a low-light level videocamera was mounted above the toy.

A number of different analyses were conducted on the head turning of the babies. In order to compare these results with those of the first experiment, naive judges again classified the infant's first major head turn (> 10°) as being in 1 of 4 90° sectors. The sector boundaries were defined by a clear plastic overlay, containing two orthogonal axes, which was placed on the TV screen with the origin centered on the infant's nose at the beginning of a trial and the axes aligned so that one sector was 45° on either side of the plane containing the sound. A response in this sector was considered "correct," whereas turns into other sectors were "incorrect." Using this criterion, the percentage of correct responses for speakers in the different planes is shown in Figure 5. Because there were no significant differences in performance between speakers positioned at 30° and 60° or between sides (left-right and up-down), the data are collapsed in the figure. Clearly, performance was best for sounds in the H plane, slightly, but significantly, poorer for sounds in the OR and OL planes, and poorest for sounds in the V plane. However, there were more correct responses (50%) in the vertical plane than would be expected by chance (25%). An analysis of silent control trials revealed that there was no difference in the probability of an infant turning into any of the 4 sectors. Finally, there were no significant differences between dim-light and dark conditions.

Forbes also used a more precise measure of head orientation derived from vector geometry. The formula was based on the assumption that 4 arrowheads placed on the infant's forehead, chin, and each cheek formed the coordinates of a flat plane with the nose projecting out from this plane. The positions of the

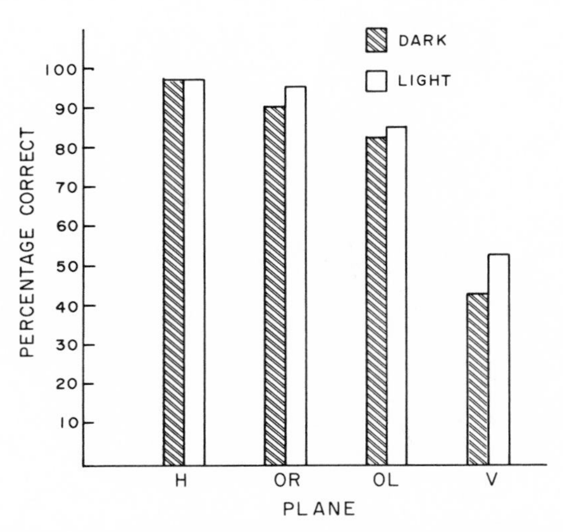

Figure 5. Percentage of head turns toward the sound in the light and dark as a function of plane: horizontal (H), vertical (V), oblique right (OR), and oblique left (OL).

arrowheads on the baby's face are shown in Figure 6A. The vertical and horizontal extents at the end of the first head movement were estimated by placing a clear overlay with an intersecting X- and Y- axis on the video screen so that they joined the points on the chin-forehead and L-R cheeks as shown in Figure 6B. Then, the distance from the midpoint of the arrowhead on the nose to each axis was measured using dial calipers and converted into degrees on the vertical and horizontal planes.

The results for turns in the correct vertical and horizontal directions are summarized in Figure 7B. The projected speaker locations at 30° and 60° from center are represented by open and solid symbols, respectively. Speaker positions are represented on the H axis by squares, the V axis by circles, and the OR/OL axes by triangles. The symbols joined by dotted lines are the averaged vertical and horizontal extents of the infants' actual head rotations toward the 30° and 60° speakers located on each plane. Only the upper auditory field is pictured; all of the data were included, but because there were no differences between sides, they have been collapsed to simplify the figure. For example, when sound came from speakers 30° and 60° along the horizontal plane, infants pointed their noses approximately 30° and 55°, respectively, along this plane. There was almost no vertical component in their head movements. By contrast, when sound came from speakers along the vertical plane, infants rotated their heads about 20° toward the 30° speakers and significantly further, but only 30° toward the 60° speakers. They also took significantly longer, just as in the first study, to initiate a head turn (> 5°) toward sounds on the vertical plane (average latency of 2.5 sec) than toward sounds on the oblique and horizontal planes (average latency of 1.1 sec).

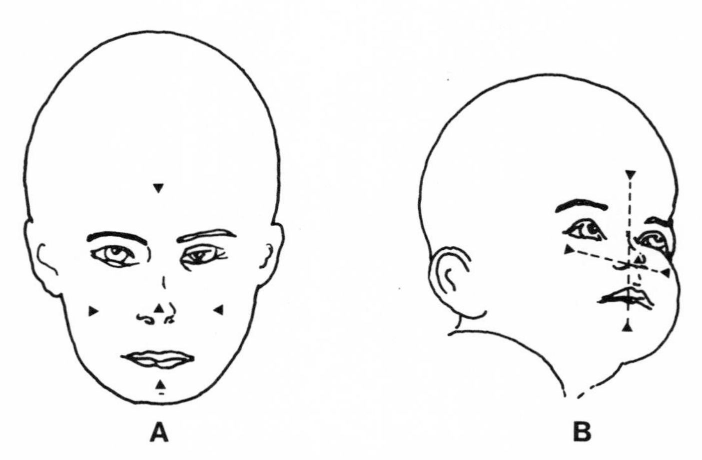

Figure 6. Drawing of the infant's face illustrating the position of the arrowheads used to calculate the vertical and horizontal extent of the head rotation.

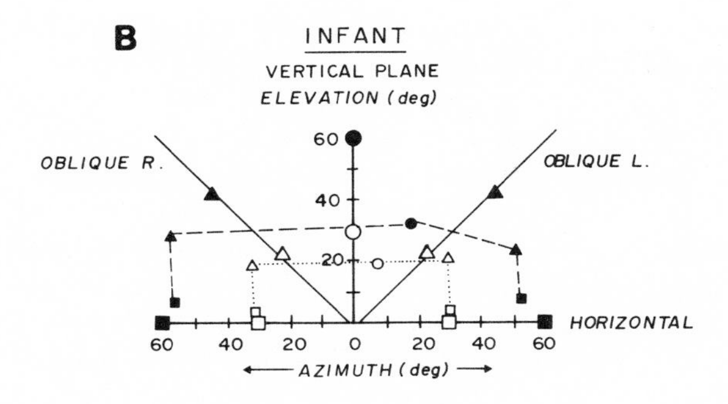

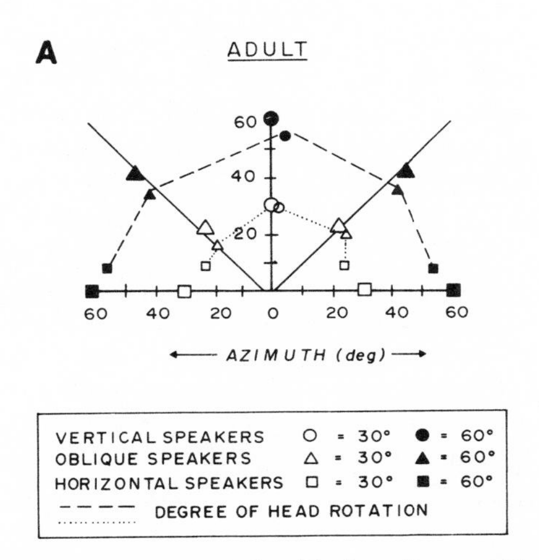

Figure 7. Two-dimensional plot of the vertical and horizontal extent of head rotation (points connected by dashed lines) of (B) 4½-month-olds and (A) adults to speakers in the horizontal, vertical, and oblique planes.

For purposes of comparison, and to validate our measures, 5 adults were also tested using the same procedure as that used with the infants, except that they were tested in the dark and were told to orient toward the sound source. Their head rotations toward speakers in different planes, calculated as with the infants, are shown in Figure 7A. In the horizontal plane, the old infants' performance matches that of the adults, whereas in the vertical plane, the adults rotate their heads farther and more closely in line with the sound source.

Taken together, these two studies provide compelling evidence that reliable localization responses can be obtained when sounds are located in the vertical, as well as in the horizontal, plane by infants as young as 4½ months of age. However, when sounds are in the two different planes, their performance is easily distinguished. Vertical plane localization is always slower and less accurate than horizontal plane localization.

Vertical Localization Responses of Newborns

Having shown that reliable vertical localization responses could be elicited from 4½-month-olds, we decided to test neonates. We used the same procedure in our first vertical localization experiment, with the following exceptions: the newborns were held semisupine as described earlier; they had to remain quiet and alert for at least three 5-trial blocks to be included (16 of 21 infants met this criterion); the neonate's head was centered between trials by the holder's face, rather than a mechanical toy; and trials lasted for a maximum of 20 sec if there was no response. As before, the holders were unaware of the stimulus location. During the trial the baby was tested in almost totally dark conditions. Head turns were judged, by naive observers, to fall into either "right," "left," "up," or "down" categories. From past experience, we knew that newborns rarely moved their heads directly upward along the vertical median plane under any circumstances. Thus, in order to provide a description of their vertical head rotations, a response was classified as "up" or "down" even if a large left or right turn also occurred, so long as there was a clear vertical component, defined as falling outside a 30° sector surrounding the horizontal plane.

The results from this analysis are shown in Figure 3B. As expected, reliable orientation toward sounds on the left and right occured. To our surprise, however, newborns turned with almost equal frequency toward sounds above the horizon (15 of the 16 babies had at least 1 "up" response in this condition). Such a response was much less frequent when the sound was below or behind the baby's line of sight and almost never occurred when it was to the left or right. On an individual basis, in the horizontal and vertical "up" sound conditions, 14 of the 16 infants turned into one of the sectors more often than the others and 13 of them favored the sector containing the sound. Also, the extent of turning was never as great toward sounds on the vertical plane as for those on the horizontal plane. Finally, few "down" responses occurred in any condition, including trials when sound was below the horizon; babies performed in a similar manner when sounds were in the lower field and behind their heads, distributing their responses relatively evenly in the other three directions.

The results of these three studies indicate that although orientation toward sounds off the horizontal plane is not very dramatic in newborns, it does occur at above chance levels, and by 4½ months of age, babies will respond correctly on about 70% of the trials. Thus, at both ages, infants can demonstrate rudimentary appreciation of the elevation of sound sources. No vertical sound localization literature exists for newborns. However, Mendelson and Haith (1976), who conducted a series of studies concerning the effect of sound on the newborn's scanning of visual targets and blank fields, in darkness, repeatedly found that their infants fixated high in the central field at the beginning of sound presentations. Their sound source was 60° above the infant's line of sight in the median plane. In spite of this, they cautiously suggested that these vertical eye movements reflected a general orienting reflex, rather than localization, but did not vary sound location in order to rule out the latter interpretation.

Our results do contrast with the few studies on vertical localization responses of older infants, which suggests that such responses are never seen until a much later age. Of course, any number of procedural differences between the various studies may have contributed to this discrepancy. The most obvious ones are our use of video recordings to analyze the response more precisely and testing under dim or dark conditions to eliminate visual interference.

Auditory Localization of Infants with Auditory or Visual Handicaps

Infants with Hearing Impairments

An obvious question of both theoretical and clinical relevance concerns the utility of the auditory localization task in the early detection of hearing disorders. We have tested several newborns with clearly established hearing impairment. Two babies had severe bilateral deafness, and neither of them turned toward off-centered sounds, even to intense noise (90 to 110 dB). As noted above, we found that approximately 25% of a large sample of newborns did not turn reliably toward sounds yet did not appear to have hearing deficits. It is nevertheless reasonable to assume that the localization test might be useful in detecting infants with unilateral hearing losses. We (Muir & Harris, 1980) have tested two babies with clear *a priori* evidence of such a loss. They both had congenital, unilateral atresia and microtia; on their right side, the pinna was absent and the external ear canal was blocked by an unresolved meatal plug. Frequently, middle-ear anomalies and severe hearing loss are associated with such complete malformations (see Jaffe, 1977). The degree of hearing loss associated with the damaged side of these two infants is not known at this time, but even with intact middle-ear function, there would be a significant interaural difference in the intensity and spectral characteristics of sounds. Thus, we expected some decrement in performance on the localization task to be readily apparent. If, for example, newborns determine the direction of sounds along the horizontal plane using interaural amplitude difference cues, these atresia infants should always turn toward the louder side, in the direction of their normal ear.

These two babies were tested extensively, beginning at birth, using our standard methods described above. Also, in all but one test session, the rattle sound was presented through speakers. The longitudinal test results for each baby (SL and KG) are summarized in Figure 8. The percentage of correct turns, out of the total number of trials on which a head turn occurred, is plotted as a function of each baby's age. The total number of trials given at each age is bracketed within the bars; the data collected during the newborn period are collapsed across several sessions because responding was almost identical on each occasion. As shown in the figure, SL, who was premature, performed as well on his preterm test sessions as he did at term, replicating our earlier work, while KG, who was a fullterm baby, was even more accurate in turning toward sounds during her newborn period. Both babies' performances dropped to chance levels when they were 3 months old and rose again at the 6- and 8-month testing to about 90% correct, similar to the accuracy of babies with normal hearing.

We are not suggesting that these two atresia infants had normal localization abilities. Both babies were also tested at 8 months of age using the Forbes and Muir (1981) procedure described above, with the speakers placed at 30° and 60° on the rotatable steel hoop. Their videotaped performance was scored using a criterion whereby a correct head turn had to be within the 90° sector surrounding the plane containing the sound and a turn into any of the other 3 sectors was

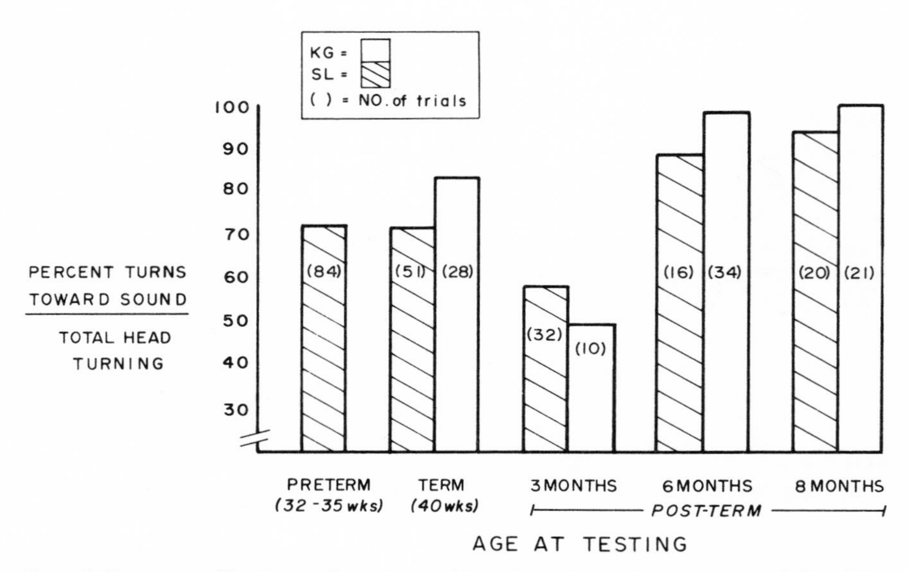

Figure 8. Percentage of head turns directed toward lateral sounds at various ages by two infants (KG and SL) with unilateral hearing losses.

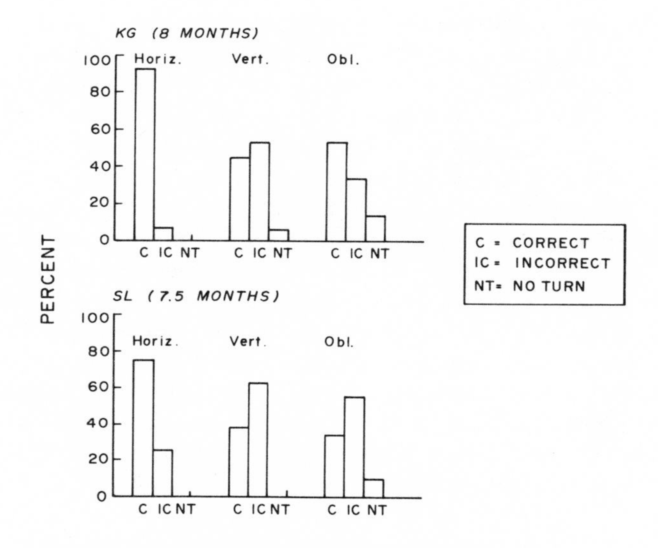

Figure 9. Percentage of trials on which infants with unilateral hearing loss directed their head turns toward or away from speakers in the horizontal, vertical, and oblique planes.

incorrect. The results of this analysis are shown in Figure 9 for each subject when sounds were presented in the H, V, and the two oblique planes, combined. Clearly, performance with stimuli on the horizontal plane is normal, but the responding drops to chance levels when the sound is in the vertical and oblique planes, and is much worse than that exhibited by normal 4½-month-olds.

These findings emphasize several points. First, infants with unilateral hearing losses can exhibit the same U-shaped developmental function for auditory localization as normal infants. Thus, one must question the diagnostic value of localization tests for early detection of auditory deficits. At the same time, this result indicates the robustness of the U-shaped function. The second point concerns the directional cues these atresia infants might be using. Both babies were able to locate sounds on the horizontal plane at birth and again at 6 months of age, in spite of the constant distortion in interaural intensity differences. These positive results may signify that they have a functioning middle ear on the damaged side and that they used interaural time difference cues, which would remain unaffected by the natural earplug, to detect the sound's direction on the horizontal plane. Of course, we do not know the cues that infants with normal hearing use to localize these sounds.

Bundy (1980) does provide evidence that very young infants can detect a phenomenal shift in sound location following habituation to an auditory-visual display. His 8-week-olds responded to change in interaural arrival time, but not intensity difference, whereas his 16-week-olds responded to changes in both cues. Possible alternative cues for our atresia infants are monaural head shadow, shoulder reflection, and/or pinna cues. However, the necessity of *a priori* information about the sound spectrum for unambiguous localization with these cues (Searle *et al.*, 1976) makes such a possibility unlikely.

The invulnerability of localization was the case for horizontal but not for vertical localization in the atresia infants. Their failure to perform as well as most normal 4½-month-olds with sounds above and below the horizontal plane suggests that distortions in pinna cues can destroy an infant's ability to determine a sound's elevation. These infants may need extensive experience in order to ignore their faulty pinna cues and to associate systematic shifts in a sound's frequency spectrum with its angle of elevation. Perhaps poor performance on vertical localization tasks may be a good predictor of certain types of congenital hearing losses, although such a relation needs to be established with subjects with unilateral deficits but normal external ears.

Auditory-Spatial Responses of a Blind Infant

As mentioned earlier, Bower has asserted that intersensory perceptual coordination is present at birth, or shortly thereafter, and does not require learning. In its extreme form, his theory predicts that very young infants experience "perfect intersensory substitutability" (Bower, 1978, p. 94). Bower (1977) claimed that one infant who wore an ultrasonic spatial sensor had "no problem" using it to explore his environment when he was 4 months old. Leslie Kay invented this aid which transmits ultrasonic pulses through a wide beam (60°) and senses the reflections from objects by 2 receivers spaced a small distance apart. These receivers translate the reflections into audible signals, which are fed to each ear through small speakers. The interaural amplitude difference serves as the cue for object position on the horizontal plane. Object distance is coded as a change in frequency (high/far and low/near), object size is coded in terms of intensity, and shape and texture are coded by the signal's timbre (see Goldstein & Wiener, 1981, for an acoustic analysis of the sonic guide). Bower reported that his infant subject, immediately upon hearing these signals from the aid, defended himself from looming objects, reached accurately for nonthreatening objects, and exhibited social reponses as the mother moved in and out of the aid's field of operation. Bower reported that the infant became even more proficient with experience; by 8 months of age he searched for and found objects that disappeared behind screens.

Such a performance by a blind baby is surprising given that normal 4-month-olds are just beginning to orient their heads again toward invisible sounds, and their responses to sounds in space off the horizontal plane are, at best, inaccurate and hard to elicit. Furthermore, blind infants tend to be delayed in almost every motor milestone, including those that do not depend exclusively on vision, such as reaching for sounds (e.g., Adelson & Fraiberg, 1974). Fraiberg (1977) reported that blind infants, following specific training, began to reach toward sounding objects at 8 to 10 months of age. Thus, we were anxious to repeat Bower's observations.

We (Dodwell, Harris, & Muir, 1979; Harris & Muir, 1980) have been studying a congenitally blind child's use of a sonar aid for several years. Our subject (NV) was premature (born at 28-weeks gestation) and contracted retrolental fibroplasia, which left her totally blind due to bilateral retinal detachment. Her parents were anxious to participate in any intervention program that might help NV to cope with her disability. She was assessed in our follow-up study of high-risk infants and, except for her blindness, had no other physical (or mental) handicaps.

Using our standard procedure with real rattles, NV's ability to orient her head toward sounds was assessed in her home at 40-weeks-gestational age and again at 3, 6, and 11 months of age (postterm), as well as at weekly intervals between 6 and 8½ months of age. At all ages, she received at least 8 rattle trials and additional trials with voices and novel sound stimuli. The percentage of major head turns toward the sounds, out of the total number of turns, is plotted as a function of age in Figure 10. Clearly, NV did not have a normal developmental function for auditory localization. At term, she failed to localize sounds reliably, turning equally often toward and away from the stimuli; at 3 months of age, she did not turn her head at all in the presence of sounds; and at 6 months, she turned once in the wrong direction. Clearly, she did not appear to treat sounding objects in the same manner as sighted infants treated visible objects.

When NV was 6 months old, Leslie Kay loaned us the sonar aid illustrated in Figure 11, which was designed for infants (described in detail by Kay & Strelow, 1977; Strelow & Boys, 1979). Following Bower's lead, we placed this aid on NV's head and videorecorded several hours of her behavior with the aid turned on and with it turned off. The aid's signal was recorded on one of the audio channels of the videorecorder. Throughout testing, the aid's signals were also audible to the experimenter through a set of earphones.

When the aid was first put on NV's head, she fussed, but when it was switched on, she usually quieted and appeared to be listening. Although she showed this reliable alerting response to the audio signals, she never turned her head or reached toward objects producing the signals. Although we were disappointed with the initial results, they were not surprising, given NV's failure to turn toward or reach out for naturally sounding objects. In spite of this, NV's parents were enthusiastic about the aid's possibilities and agreed to help prepare NV for using the aid. Thus, the use of the aid was postponed for several months

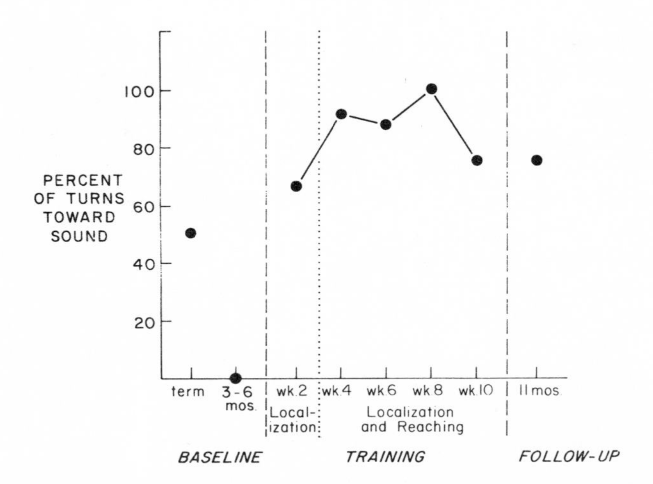

Figure 10. Percentage of head turns directed toward sound by a congenitally blind infant (NV) at different ages. Baseline refers to testing carried out from birth to 6 months of age. The data plotted between the vertical dashed lines represent 2-week blocks of performance during the 10-week-training period beginning at 6 months of age. The vertical dotted line separates the time when only head turns were reinforced, from the time when both turns and reaches were reinforced. The last point is performance on the posttest 2½ months after formal training was terminated.

and an exercise program was instituted. Beginning when she was 6 months old, her parents trained her daily with social reinforcement, hugs and kisses, to turn toward sounds for the first 2 weeks, and then to turn, reach, and grasp for them for another 8 weeks. The results of weekly tests, collapsed over 2-week periods, during this 10-week training phase are shown in Figure 10 along with the results of a posttest 1½ months after the end of the training period. Although head turning increased in frequency over earlier levels (70% of the trials), NV turned away from the sound source almost as often as she turned toward it during the first 2 weeks of training. By the third and fourth weeks, however, she turned 90% of the time in the correct direction, when she turned at all (65% of the trials). By the eighth week of training, she turned in the correct direction on all trials. Her average latency to complete a response decreased during training from 4.6 sec to 2 sec. NV's arm movements, both ipsilateral and contralateral to the sound, occurred with approximately equal frequency, on about 80% of the trials during the first two weeks of localization training and continued at this high, undifferentiated level for several weeks after differential reinforcement for correct responses was initiated. By the eighth training week, NV was reaching primarily with the arm closest to the sound. Thus, by the time NV was 7 to 8 months old, she was capable of locating near objects by their sounds.

As a final preparation to wearing the aid, NV wore a velcro headband on her head intermittently during her eighth month. Her initial response to the aid the second time she wore it, when she was 9 months old, differed markedly from that 3 months earlier. When the aid was first turned on, NV heard the signal reflected from the carpet and she reached out and touched various spots on the floor for several seconds; then, finding nothing, she reached up and touched the sonar aid speaker next to her left ear. Thus, she appeared to regard the aid's signal as being related to the presence of an object in space and not finding anything in a brief search, she raised her hand to the actual source. NV gave a further indication that she related the signal to an external source by reaching (without contact) on 50% of the trials when a silent object was brought into the aid's field, but not on similar trials with the aid turned off. Finally, she showed an initial sensitivity to the binaural direction code by successfully following, with her head, a silent object moving left and right on 7 of 10 trials and exhibiting no such tracking of silent objects with the aid turned off.

After this initial responsiveness, NV's spontaneous activity during free play with the aid on was little different from when it was turned off. Although she wore the aid for several hours a day, she did not appear to use the information to

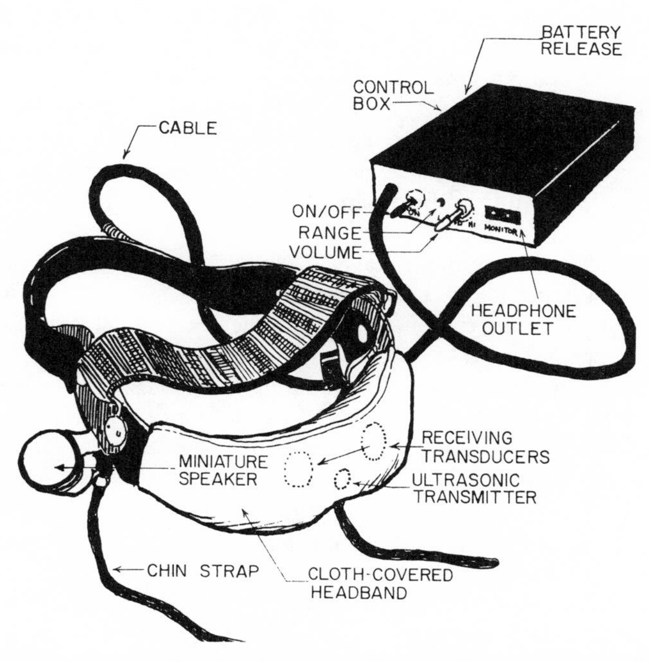

Figure 11. An illustration of Kay's ultra-sonic binaural spatial sensor for blind infants. The headband has adjustable straps and the control box/power supply is carried in a backpack.

explore her hand movements or to search for silent objects that she dropped within reach. Furthermore, she never anticipated silent looming objects with defensive gestures, reaching, or changes in facial expression. She could be trained, however, to anticipate looming objects. After approximately a month of daily practice with approaching spoons of food, she made anticipatory mouth openings on 67% of the trials with the aid turned on and only 29% of the time with it turned off. However, this behavior was quite variable from one feeding to the next and did not stabilize until she was over 12 months old. NV's ability to capture silent toys presented at various positions in the aid's field was assessed weekly for the first 6 months that she wore the aid. It was several months before she was able to successfully contact the toys within reach using the aid. Initially, she contacted silent toys about as often with the aid switched on as off; then she stopped reaching when the aid was off and succeeded in grasping them on about 40% to 50% of the trials with the aid turned on.

Clearly, NV could use the aid's signal to locate and reach for objects and to anticipate collisions with approaching objects. But, as was typical of each testing and training program we introduced, she eventually became bored with our games, and in the case of reaching, after a few months of aid use, she frequently refused to reach or simply swept her arms out in front of her until she established contact. The same was true when she began to walk. At 20 months of age, she reliably stopped just before contacting an obstacle or wall, on 80% of trials, as she walked along holding someone's hand. When she strolled out on her own for the first time, her parents reported that she was wearing the aid and by 21 months of age she stopped in anticipation of a collision on about 60% of trials with the aid on and only 10% with it off. However, once she mastered this task, she frequently would playfully collide with objects, as though this were part of her exploration strategy.

To conclude, each milestone that NV achieved with the aid was preceded by a lengthy period of training diligently carried out by her parents. Even the brief, apparently natural spatial responses made to the aid's signals when she was 9 months old were preceded by considerable training to encourage her to find audible objects. Perhaps the most disappointing result was that, even after several years of experience, NV never seemed to spend much time using the aid's signal to explore her environment during training or testing. In summary, although a number of skills using the sound cues provided by the sonar aid could be trained, NV never responded in the manner of Bower's subject. Perhaps his infant had some residual vision and the aid's signal provided an attentional supplement to the minimal visual cues.

Our observations of NV's development are in line with those of Fraiberg (1977) and others, who find delays in acquiring most early motor skills. Presumably, any attention these infants might initially give to distant, interesting sounds is soon extinguished due to a lack of immediate or even delayed reinforcement.

Indeed, the premobile blind child's representation of space may be insufficient to permit the learning of associations between signal frequency and distance. This might explain why, in spite of the fact that NV could respond to directional information at 9 months of age, this skill was rarely used.

Summary and Conclusions

If we assume that the infant's functional representation of auditory space is reflected in orientation behavior toward sounds in different spatial locations, the following summary picture is obtained. Newborns appear to place sounds in a long, narrow band along, and just above, the horizontal plane. This "space" is either ignored or blurred when the baby is about 2 months old and reappears fairly abruptly in a more accurate form at about 4½ months of age. At this time, the space is well defined, perhaps adultlike, along the horizontal axis and while its reach, both above and below this axis, the vertical dimension is still narrow, compared to the adult's. Although a complete 2-dimensional developmental function has yet to be determined, the vertical dimension presumably expands during the latter part of the first year, as suggested by Northern and Downs' (1978) clinical observations. One might argue that newborns do not align their heads as far in the vertical direction simply because of poor motor control, but this explanation does not apply to 4½-month-olds who are adept at turning to face visual targets 60° above their line of sight. Another point to note is that the vertical localization responses of 4½-month-olds are, in several respects, very similar to their horizontal responses prior to this age; they tend to turn slowly and in the correct direction on only about 70% of the trials and to underestimate the distance from midline. Perhaps precise responses aligned with the actual elevation of sounds also emerge abruptly at some later age. These findings are compatible with the view that auditory space is complex, compared with vision, and constructed from a variety of cues, and that the formation of an adultlike 2-dimensional representation requires time and experience.

In a larger context, our findings support different aspects of both differentiation and integration theories of perceptual development, as they are proposed by Bower and Piaget, respectively. The fact that newborns will turn toward sounds along the horizontal and, to some extent, the vertical planes at birth is in line with Bower's idea of an innate, spatial unity of the senses. That infants also orient toward off-centered visual and tactile stimulation adds further weight to his theory that they may be operating in a united, multimodal space. But no direct evidence on the exact nature of such a space actually exists. Clearly, Piaget was wrong when he asserted that infants would not look toward sounds during their first month of life. But he may have been correct when he classified newborn responses as simple reflexes. At this age they may be simply turning toward lateral stimulation in a tropistic manner, much like a worm orients toward the dark. The consistent results showing that spatial responding to sounds diminishes during the second month of life, along with newborn reflexes, make this latter interpretation seem plausible. Our finding that the timing of this drop in localization responding is related to the baby's conceptional, rather than chronological, age adds further support to Piaget's emphasis on maturational factors in sensorimotor coordination.

Neither theory seems to account readily for the persistence of diminished localization between the second and fourth month of life. Bower might say that this is when differentiation begins, but he had already implied that multimodal space remains unified throughout most of the first year of life. Piaget believed that infants first coordinate their auditory and visual localization schemas beginning at 2 months, as they visually explore their environment. They may do so, but the evidence seems weak. In spite of the 2-month-old's apparent ability to respond to localization cues (e.g., Bundy, 1980), many investigators find that the position of a sound is relatively unimportant to infants of this age. Aronson and Rosenbloom (1971) did suggest that 2-month-olds became upset when their mothers' voices were dislocated from their mothers' faces. But other investigators have not been able to reproduce this phenomenon (e.g., Field *et al.*, 1979; Lewis & Hurowitz, 1977; McGurk & Lewis, 1974). Some weak, but reliable, spatial influence of sound has been demonstrated under certain circumstances. For example, McGurk and MacDonald (1978) found that if babies were presented with two identical visual stimuli and sound was played from behind one of the displays, 3-month-olds would spend slightly more time looking at the side with sound. Also, Bull (1981) has shown that 4-month-olds' visual search for the reappearance of a moving visual target which temporarily disappears behind an occluding screen is facilitated by adding a continuous sound to the visual target. However, the effect was slight and there was no evidence to suggest that spatially discordant sounds, either stationary or moving in the opposite direction, detracted from the infants' tracking of, or search for, visual targets.

Because infants appear to be disturbed so little by a lack of spatial contiguity between related sights and sounds Spelke (see Spelke, 1979, for a review) was able to invent an elegant paradigm for demonstrating that infants as young as 4 months of age associate particular auditory patterns with appropriate visual patterns. For example, she presented them with two adjacent film displays, one of a woman playing peek-a-boo and the other of a baton striking a tambourine. Infants tended to look toward the film that was associated with the appropriate sound track presented between the two visual displays. Perhaps the major developmental task after the second month of life is to relate common auditory-visual patterns irrespective of the sound's location.

The role played by vision in directing and maintaining auditory localization responses during the first 6 months of life is unclear. Both newborns and 4½-month-olds do not have to see something at the position of auditory targets to continue turning toward them for 10 to 30 trials (e.g., our testing in the dark). Indeed, investigators wishing to employ localization responses for clinical applications and to determine sensory thresholds have been frustrated in their attempts to visually reinforce turns toward sounds if infants were less than 6 months old (e.g., Field, 1981; Moore, Thompson, & Thompson, 1975; Schneider, Trehub, & Bull, 1979). On the other hand, some form of reinforcement appears to be necessary to lure infants toward sounding objects. Our blind subject failed to show localization behavior on all of her tests until she was over 6 months old, and then only after extensive training.

Our observations of NV's inattention to auditory spatial cues are in agreement with those of Fraiberg (1977) and others, who found that sounds of natural objects, even prized toys, did not begin to elicit spontaneous searching by blind infants until the latter half of their first year. Indeed, although sighted infants automatically orient toward sounds (Muir *et al.*, 1979) and temporarily reach for sounds in the dark (Wishart, Bower, & Dunkeld, 1978) when they are 4 to 5 months old, they too fail to search reliably for desired objects, which disappear from sight, based on sound cues alone until the end of their first year (e.g., Freedman, Fox-Kolenda, Margileth, & Miller, 1969; Uzgiris & Benson, 1980). Given these findings, we question Bower's contention that blind infants, during their first 6 months of life, have the ability to see with their ears using the sonar aid. It is still too early to ascertain the effect of early experience with the sonar aid on NV's later spatial abilities, but one result is clear. In spite of major efforts by her parents and extensive exposure to the aid, NV never displayed the spatial awareness of the infant studied by Bower. She did generalize from her trained skill of reaching to natural sound cues to reaching "based on" cues provided by the sonar aid. However, she never exhibited dramatic examples of exploring the environment for silent objects or playing peek-a-boo games, either during her first exposure to the aid at 6 months of age or later between 9 and 12 months of age. Her apparent lack of understanding of the aid's signal seems reasonable given the complex coding of object size, distance, and directional information. If sighted infants with visual representation of space derive minimal functional information about objects from sound cues, it would follow that blind infants would not immediately construct objects in space from the abstract auditory cues provided by the sonar aid.

Even though our blind infant responded to the aid's direction cue and eventually to the distance cue, anticipating looming objects, she may not have been able to relate these signals to positions in an abstract space. Perhaps the blind child's early representation of space is too limited to permit associations between signal frequency and distance. Of course, the results with one infant hardly constitute conclusive evidence against Bower's theory of intermodal sensory equivalence, but healthy scepticism at this point seems appropriate.

ACKNOWLEDGMENTS

The contributions of H. Killen and B. Pater (nurses who held the newborns), Kate Bolen and Frances Whalley (research assistants who helped test and score the infants), and the parents and their infants who participated in this research are gratefully acknowledged. Others who made significant contributions were Dr. B. Shelton, who analyzed the tape recorded rattle sounds at the Harvard Laboratory of Psychophysics, and Dr. J. C. Hogarth, Statlab, Queen's University, who developed the formula for calculating the degree of head rotation used in Forbes (1981). Special thanks are due to the parents of the blind infant (NV) for their earnest cooperation over the years and to Drs. J. Field, F. Wilkinson, K. Humphrey, and D. DiFranco for their important contributions during the lengthy training and evaluation program for NV's use of the sonar aid. Finally, Bob Pilon and Ann Muir provided a helpful critical review of the manuscript and Lorraine Adams produced the graphics.

References

Adelson, E., & Fraiberg, S. Gross motor development in infants blind from birth. *Child Development,* 1974, *45,* 114–126.

Alegria, J., & Noirot, E. Neonate orientation behavior towards human voice. *International Journal of Behavioral Development,* 1978, *1,* 291–312.

Aronson, E., & Rosenbloom, S. Space perception in early infancy: Perception within a common auditory-visual space. *Science,* 1971, *172,* 1161–1163.

Bayley, N. *Bayley Scales of Infant Development.* New York: Psychological Corporation, 1969.

Bower, T. G. R. *Development in infancy.* San Francisco: Freeman, 1974.

Bower, T. G. R. Blind babies see with their ears. *New Scientist,* 1977, *73,* 255–257.

Bower, T. G. R. Perceptual development: Object and space. In E. C. Carterette & M. P. Friedman (Eds.), *Handbook of perception* (Vol. VIII). New York: Academic Press, 1978.

Bower, T. G. R. The origins of meaning in perceptual development. In A. D. Pick (Ed.), *Perception and its development: A tribute to Eleanor J. Gibson.* Hillsdale, N.J.: Erlbaum, 1979.

Brazelton, T. B. *Neonatal Behavior Assessment Scale.* London: Spastics International Medical Publications, 1973.

Bull, D. H. *Infants' visual tracking of sights and sounds.* Unpublished doctoral dissertation, University of Toronto, 1981.

Bundy, R. S. Discrimination of sound localization cues in young infants. *Child Development,* 1980, *51,* 292–294.

Butterworth, G. The origins of auditory-visual perception and visual proprioception in human development. In R. D. Walk & H. L. Pick (Eds.), *Intersensory perception and sensory integration.* New York: Plenum Press, 1981.

Coghill, G. E. *Anatomy and the problem of behavior.* London: Cambridge University Press, 1929.

Chun, R. W. M., Pawsat, R., & Forster, F. M. Sound localization in infancy. *Journal of Nervous and Mental Disorders,* 1960, *130,* 472–476.

Clifton, R. K., Morrongiello, B. A., Kulig, J. W. & Dowd, J. M. Newborns' orientation toward sound: Possible implications for cortical development. *Child Development,* 1981, *52,* 833–838.

Crassini, B., & Broerse, J. Auditory-visual integration in neonates: A signal detection analysis. *Journal of Experimental Child Psychology,* 1980, *29,* 144–155.

Dayton, G. O., & Jones, M. H. Analysis of characteristics of fixation reflex in infants by use of direct current electroculography. *Neurology,* 1964, *14,* 1152–1156.

Dodwell, P. C., Harris, L., & Muir, D. W. *Acceleration of cognitive growth: The case of a blind infant.* Paper presented at the Canadian Psychological Association, Quebec, June, 1979.

Drillien, C. M., & Drummond, M. B. (Eds.), *Neurodevelopmental problems in early childhood: Assessment and management.* London: Blackwell, 1977.

Field, J. Looking and listening in the first year of infancy. In A. R. Nesdale, R. Pratt, R. Grieve, J. Field, D. Illingworth, & J. Hogben (Eds.), *Advances in child development: Theory and research.* Belmont, Western Australia: Quality Press, 1981.

Field, J., DiFranco, D., Dodwell, P., & Muir, D. Auditory-visual coordination of 2½-month-old infants. *Infant Behavior and Development,* 1979, *2,* 113–122.

Field, J., Muir, D., Pilon, R., Sinclair, M., & Dodwell, P. Infants' orientation to lateral sounds from birth to three months. *Child Development,* 1980, *51,* 295–298.

Fisher-Fay, A. *Auditory-visual conflict in newborn infants.* Unpublished honours thesis, Queen's University, 1981.

Forbes, B. *Orientation differences in sound localization abilities of 4 to 5 month old infants.* Unpublished master's thesis, Queen's University, 1981.

Forbes, B., & Muir, D. W. *Orientation differences in sound localization abilities of four month old infants.* Paper presented at the Canadian Psychological Association, Toronto, Ontario, June, 1981.

Forbes, B., Abraham, W., & Muir, D. W. *The accuracy of newborn auditory localization.* Paper presented at the Canadian Psychological Association, Quebec, June, 1979.

Fraiberg, S. *Insights from the blind.* New York: Basic Books, 1977.

Freedman, D. A., Fox-Kolenda, B. J., Margileth, D. A., & Miller, D. H. The development of the use of sound as a guide to affective and cognitive behavior—A two phase process. *Child Development,* 1969, *40,* 1099–1105.

Goldstein, B. A., & Wiener, W. R. Acoustic analysis of the sonic guide. *Journal of the Acoustical Society of America,* 1981, *70,* 313–320.

Haith, M. M. Visual competence in early infancy. In R. Held, H. Leibowitz, & H. L. Teuber (Eds.), *Handbook of sensory physiology* (Vol. 8). Berlin: Springer-Verlag, 1978.

Hammond, J. Hearing and response in the newborn. *Developmental Medicine and Child Neurology,* 1970, *13,* 3–5.

Harris, L. & Muir, D. *Auditory-motor and sonar aid training with a blind infant.* Paper presented at the Ontario Psychological Association, Toronto, Ontario, June, 1980.

Hecox, K. Electro-physiological correlates of human auditory development. In L. B. Cohen & P. Salapatek (Eds.), *Infant perception: From sensation to cognition.* New York: Academic Press, 1975.

Jaffe, B. F. Atresia: Isolated and associated anomalies. In B. F. Jaffe (Ed.), *Hearing loss in children.* London: University Park Press, 1977.

Kay, L., & Strelow, E. Blind babies need specially designed aids. *New Scientist,* 1977, *74,* 709–712.
Kurtzberg, D., Vaughan, H. E., Daum, C., Grellong, B. A., Albin, S., & Rotkin, L. Neurobehavioural performance of low-birthweight infants at 40 weeks conceptual age: Comparison with full-term infants. *Developmental Medicine and Child Neurology,* 1979, *21,* 590–607.
Lewis, M., & Hurowitz, L. *Intermodal person schema in infancy: Perception within a common auditory-visual space.* Paper presented at the Eastern Psychological Association, April 1977.
McGraw, M. B. *The neuromuscular maturation of the human infant.* New York: Columbia University Press, 1943.
McGurk, H., & Lewis, M. Space perception in early infancy: Perception within a common auditory-visual space? *Science,* 1974, *186,* 649–650.
McGurk, H., & MacDonald, J. Auditory-visual coordination in the first year of life. *International Journal of Behavioral Development,* 1978, *1,* 229–240.
Mendelson, M. J., & Haith, M. M. The relation between audition and vision in the human newborn. *Monographs of the Society for Research in Child Development,* 1976, *41* (No. 4).
Moore, J. M., Thompson, C., & Thompson, M. Auditory localization of infants as a function of reinforcement conditions. *Journal of Speech and Hearing Disorders,* 1975, *40,* 29–34.
Muir, D. W. The development of human auditory localization in infancy. In W. Gatehouse (Ed.), *Sound localization: Theory and application.* Groton, Conn.: Amphora, 1982.
Muir, D. W., & Field, J. Newborn infants orient to sounds. *Child Development,* 1979, *50,* 431–436.
Muir, D. W., & Harris, L. *Localization responses in normal and high-risk infants.* Paper presented at the International Conference on Infant Studies, New Haven, Connecticut, April, 1980.
Muir, D. W., Abraham, W., Forbes, B., & Harris, L. The ontogenesis of an auditory localization response from birth to four months of age. *Canadian Journal of Psychology,* 1979, *33,* 320–333.
Northern, J. L., & Downs, M. P. *Hearing in children.* Baltimore: Waverly Press, 1978.
Piaget, J. *The origins of intelligence in children.* New York: Norton, 1952.
Pick, H. L., Yonas, A., & Rieser, J. Spatial reference systems in perceptual development. In M. H. Bornstein & W. Kessen (Eds.), *Psychological development from infancy.* Hillsdale, N.J.: Erlbaum, 1979.
Rieser, J., Yonas, A., & Wikner, K. Radial localization of odors by human newborns. *Child Development,* 1976, *47,* 856–859.
Searle, C. L., Braida, L. D., Davis, M. F., & Colburn, H. S. Model for auditory localization. *Journal of the Acoustical Society of America,* 1976, *60,* 1164–1175.
Schneider, B. A., Trehub, S. E., & Bull, D. The development of basic auditory processes in infants. *Canadian Journal of Psychology,* 1979, *33,* 306–319.
Spelke, E. S. Exploring audible and visible events in infancy. In A. D. Pick (Ed.), *Perception and its development: A tribute to Eleanor J. Gibson.* Hillsdale, N.J.: Erlbaum, 1979.
Strelow, E. R., & Boys, J. T. The Canterbury Child's Aid: A binaural spatial sensor for research with blind children. *Journal of Visual Impairment and Blindness,* 1979, *5,* 179–184.
Turkewitz, G., Gordon, E. W., & Birch, H. G. Head turning in the human neonate: Effect of prandial condition and lateral preference. *Journal of Comparative and Physiological Psychology,* 1965, *59,* 189–192.
Turkewitz, G., Birch, H. G., Moreau, T., Levy, L., & Cornwell, A. C. Effect of intensity of auditory stimulation on directional eye movements in the human neonate. *Animal Behavior,* 1966, *14,* 93–101.
Turner, S., & Macfarlane, A. Localization of human speech by the newborn baby and the effects of pethidine ('meperidine'). *Developmental Medicine and Child Neurology,* 1978, *20,* 727–734.
Uzgiris, I. C., & Benson, J. *Infant's use of sound in search for objects.* Paper presented at the International Conference on Infant Studies, New Haven, Conn., April 1980.

Uzgiris, I. C., & Hunt, J. McV. *Assessment in infancy*. Chicago: University of Illinois Press, 1975.

Watrous, B. S., McConnell, F., Sitton, A. B., & Fleet, W. F. Auditory responses of infants. *Journal of Speech and Hearing Disorders,* 1975, *40,* 357–366.

Werner, H. L'unité des sens. *Journal de Psychologie,* 1934, *31,* 190–205.

Wertheimer, M. Psychomotor coordination of auditory and visual space at birth. *Science,* 1961, *134,* 1692.

Wishart, J. G., Bower, T. G. R., & Dunkeld, J. Reaching in the dark. *Perception,* 1978, *7,* 507–512.

Wolff, P. Observations on newborn infants. *Psychosomatic Medicine,* 1959, *21,* 110–118.

CHAPTER 4

The Precedence Effect

Its Implications for Developmental Questions

Rachel Keen Clifton

Department of Psychology
University of Massachusetts
Amherst, Massachusetts

The study of illusions forcibly reminds us that our experience of the world is, to a great extent, created within our own heads. Understanding illusions can help explicate the way the brain handles sensory information, which is an interesting idea traced through its long history by Coren and Girgus (1978). Developmental changes in the way illusions are perceived may, in turn, bring a better understanding of both the developing nervous system and the functioning of a mature nervous system.

For several years, we have been engaged in studies of the *precedence effect,* an auditory illusion in which the listener fails to localize a sound that is well-above threshold. To produce the precedence effect, two identical sounds are delivered with equal intensity from two spatial locations, with the onset of one leading the other by a certain amount of time. The time difference can vary between 4 and 40+ msec, depending on the acoustical characteristics of the stimuli. Beyond the critical time limit, one localizes both sounds at their true locations, but within the boundary of the precedence effect one localizes the sound solely at the leading loudspeaker. Even relatively continuous sounds, such as speech, certain types of music, or a rhythmically shaken rattle, display the precedence effect in that the continuous output from the lagging loudspeaker is not localized at its source. As long as the sound has a pulsating quality with some

This research was supported by grants HD-06753, BNS-810354, and MH-00332.

abrupt onsets, the illusion is maintained; breakdown occurs with long duration, pure tones, or other sustained, unvarying sounds (Wallach, Newman, & Rosenzweig, 1949).

The precedence effect is due to the nervous system's active suppression of the lagging sound because the time delays are well within the auditory system's capabilities for resolving or discriminating temporal differences. For example, when two clicks varying in interaural delays are delivered dichotically through earphones and adults are asked if two clicks are the "same" or "different," interaural time differences as small as 10 microseconds can be discriminated (Klump & Eady, 1956). An important aspect that should be emphasized about the precedence effect is its dependence on binaural input. Unlike the study of click pairs just described, each ear gets *two* inputs for each stimulus event. First, the leading sound arrives at the ipsilateral ear, then travels over the head to the contralateral ear a few *microseconds* later, supplying a major time cue for localizing sounds from single sources in the environment (Green, 1976). Second, several *milliseconds* later, the lagging sound arrives at the ear ipsilateral to it (i.e., the previous contralateral ear), then travels over the head to the other ear. In perceiving these pairs of inputs, the nervous system weights the first arriving pair more heavily, locates the sound at that position in space, and ignores information about location of the second pair of inputs. If monaural discrimination is tested by delivering two clicks to one ear (or both ears simultaneously with no interaural delay), adults can hear two distinct clicks with separations as small as about 3 msec (Wallach *et al.*, 1949). Below 3 msec auditory fusion is experienced, although as noted previously interaural differences can be discriminated at much shorter time delays.

The precedence effect has been described as an "echo suppression" mechanism that enhances the perception of the true location of a sound's source (Green, 1976, pp. 215–218; Mills, 1972). Ordinary rooms provide numerous reflecting surfaces that bounce sound back to the listener in a complex array of reverberations. These reverberations, if localized, would make the true source of the sound very difficult to localize. The ability to localize the spatial origin of a sound has obvious adaptive value for most organisms (Erulkar, 1972). As would be expected, cats (Whitfield, Cranford, Ravizza, & Diamond, 1972), rats (Kelly, 1974), and monkeys (Heffner, 1973) appear to respond to the precedence effect by localizing only the leading sound at the same temporal delays effective for human adults.

The impetus for pursuing developmental studies of the precedence effect came from research that implicated the auditory cortex in the perception of this illusion. In a series of studies, cats with lesions in the auditory cortex were impaired in their localization of precedence-effect sounds but retained the ability to localize sounds from a single source (Cranford, Ravizza, Diamond, & Whitfield, 1971; Cranford & Oberholtzer, 1976; Whitfield, 1978; Whitfield, Cranford, Ravizza, & Diamond, 1972; Whitfield, Diamond, Chiveralls, & Williamson, 1978). Children with epileptic foci in the temporal region showed impaired localization of precedence-effect stimuli, whereas they localized single source stimuli with virtually no errors (Hochster & Kelly, 1981). Normal children and adults localize both types of stimuli identically as do intact animals. We reasoned that at some early stage in development infants would fail to respond to precedence-effect stimuli but would respond to single-source stimuli. The comparison of these two stimuli is especially appealing because the same motor behavior (head turning toward sound) would be elicited, but the perceptual processing would be more difficult for precedence-effect stimuli.

Why would we make such a prediction? Substantial brain growth takes place in the first year of life, with the most dramatic development taking place in cortical regions (Dekaban, 1970; Yakovlev & Lecours, 1967). A perceptual task featuring two levels of stimulus difficulty, one of which is dependent upon cortical functioning, was expected to show a developmental progression. Older babies should respond equally well to both types of stimuli, whereas younger babies should respond only to single-source stimuli. Specific age predictions were difficult to make, but since even newborns turn their heads toward the sound of a rattle (Muir & Field, 1979), this seemed like an appropriate age to start comparisons.

General Methodology

In all research to be described in this chapter, two loudspeakers were located 90° off midline to the infant's right and left. For some studies, the stimuli were tape-recorded rattle sounds produced by manually shaking a hard plastic bottle, one-third filled with popcorn kernels, at the rate of 2/sec. In other studies, square-wave clicks, 3 msec in duration, were presented in click trains for several seconds. All infants were full-term, healthy babies, born without complications and experiencing normal postnatal development. Care was taken to test infants in a calm, alert state because either fussiness or drowsiness affects the response to sound. Infants' behavior was videotaped from a face-on position and scored later by two independent observers who did not know from which side the sound was presented or the type of stimulus. Direction of head turn was scored as right or left for lateral head movements of 10° to 15° or more off midline. Small head movements back and forth across the midline were scored as vacillation and were not considered head turns.

Hypothesis 1: Directional Responding toward Precedence-Effect Stimuli Will Be Slower to Develop in Infancy Than the Responding to Sounds from a Single Source

Muir and Field's data (1979) supplied a baseline comparison of newborn head turning toward sound. We replicated their stimulus (the rattle sound) and procedures with the addition of two new trial types. Precedence-effect trials were created by passing the signal through two loudspeakers with one output delayed relative to the other by 7 msec. Control trials were created by passing the signal through both loudspeakers simultaneously, a procedure that gives adults the perception of a centrally located sound halfway between the loudspeakers. Control trials checked on the possible arousal effects of the sound itself, measuring the spontaneous level of head turning in this situation. (For complete details of this study see Clifton, Morrongiello, Kulig, & Dowd, 1981.)

Although every newborn received an equal number of single-source, precedence-effect, and control trials, head turning occurred on 58% of the single-source trials, whereas this behavior was elicited on only 11% of precedence-effect and 17% of control trials. These results confirmed the prediction that greater complexity introduced by input and arranged to produce the precedence effect would disrupt head turning, which was usually observed toward lateralized sound. A concern that the particular delay used (7 msec) was not optimal for newborns' perception of the precedence effect led to a follow-up study in which the delay between loudspeaker onset was varied (Morrongiello, Clifton, & Kulig, 1982). Each newborn received an equal number of single-source trials and 5-msec, 20-msec, and 50-msec delay trials. Again the results were overwhelmingly clear: head turning on single-source trials was 46%, whereas head turning on precedence-effect trials was 4%, 3%, and 4% for 5-, 20-, and 50-msec delays, respectively. Heart rate was also measured in this study, and, interestingly enough, showed no differences among the various trial types. Heart-rate deceleration occurred to all types of stimuli, but only on trials without head turns.

Since newborns did not appear to localize precedence-effect stimuli in the way adults do, the obvious question was when would the illusion be present? At what point in development would the sound be clearly localized on the side of the leading signal? We chose to test 5- to 6-month-olds for several reasons. One has already been indicated: rapid development takes place in cortical regions between birth and 6 months. A second reason had more to do with behavioral competency, as head turning toward sound is well established in this age group. Newborn head turning toward sound is difficult to elicit, with the infant's arousal state, stimulus characteristics, and handling of the baby being critical elements affecting performance. By 5 months of age, infants spend more time in alert states and will turn readily toward a large variety of sounds. We expected equivalent responding to single-source and precedence-effect trials at this age, and were not disappointed. Five-month-olds turned quickly and accurately toward a train of clicks presented from one loudspeaker. When the click train was presented through both loudspeakers with a 7-msec delay between onsets, the infants turned toward the leading loudspeaker with no hesitation, or indeed any discernable difference from their behavior on single-source trials (Clifton, Morrongiello, & Dowd, 1984).

After establishing two end points for head orienting toward sound, it would appear simple to determine the intermediate age between birth and 6 months when transition behavior occurs, but, as will be seen, this step still eludes us. We selected the 6- to 9-week period as a likely transition age, as this period has been associated with other important changes in infant behavior (Emde, Gaensbauer, & Harmon, 1976). Using the click trains that were extremely effective in eliciting head turning in 6-month-olds, we tested 20 infants between 6 and 9 weeks of age (Clifton, Morrongiello, & Dowd, 1984). The same procedure of presenting every baby with equal numbers of single-source trials and precedence-effect trials was used, with side of presentation balanced across trials. Unfortunately, these infants did not turn their heads reliably toward *either* type of stimulus, rendering the comparison between them meaningless. Out of 80 possible single-source trials, head turning occurred on only 25% of the trials, with 30% of these turns in the contralateral direction. For the 80 precedence-effect trials, head turning occurred on 16% of total trials, with turns distributed equally toward the leading and lagging loudspeakers. When tested statistically, head turning toward the correct side did not differ from chance for either single-source or precedence-effect stimuli due to increased contralateral turns and the generally lower incidence of head turning at this age.

These differences from head turning in the newborn study were not due to the use of click trains rather than the rattle, as we originally suspected. In two longitudinal studies (Field, Muir, Pilon, Sinclair, & Dodwell, 1980; Muir, Abraham, Forbes, & Harris, 1979), infants tested between birth and 3 months with the rattle stimulus showed depressed levels of head turning at 2 months of age. (All trials were presented from a single spatial location.) Nevertheless, we made another attempt to elicit head turning in this age group using speech as the stimulus. Natural speech is an ecologically valid sound that might elicit attention in infants more readily than clicks. Speech is also more complex acoustically, and appears to contain crucial elements for eliciting responses even in newborns (Clarkson & Berg, 1983). We presented 20 infants, 6- to 9-weeks-old, with a tape-recorded female voice greeting a baby, using natural cadence and intonation of an adult talking to a baby. Again, each infant had four single-source and four precedence-effect trials. Speech stimuli were more effective than clicks: head turning occurred on 51% of single-source trials, with 83% of these turns in the correct direction. Head turning on precedence-effect trials was less (39%), with turns toward leading and lagging sounds distributed at chance levels (see Clifton *et al.*, 1984, for details). At 2 months of age, infants are still not responding to the precedence effect. Moreover, their head turns toward sound in general have decreased relative to both newborns and 6-month-olds, so that "optimal" stimuli must be used to obtain the behavior at all. This developmental change in localization behavior has been interpreted in terms of maturation of the auditory cortex (Clifton *et al.*, 1981; Field *et al.*, 1980; Muir & Clifton, in press).

At this point, I wish to comment briefly on a methodological aspect of newborn head turning toward sound. Muir and colleagues (Muir & Field, 1979; Field *et al.*, 1980) have reported ipsilateral head turns toward the rattle stimulus on about 80% of trials. In two studies from our lab (Clifton *et al.*, 1981; Morrongiello *et al.*, 1982), the incidence of head turning has varied between 45% and 56% of the single-source trials. Although many factors might have produced these differences between laboratories (e.g., obstetrical practices at different hospitals, nursery practices, experimenters' criteria for scoring head turn, etc.), we noticed that within our own laboratory, the percentage of head turning varied positively with the probability of a laterally produced sound. Our typical procedure has been to use within-subject comparisons to evaluate head turning toward different types of sound sources. Each infant would receive both single-source and precedence trials, with control trials consisting of either presentation of sound from a centrally located loudspeaker or simultaneous sound from both laterally located loudspeakers. Both types of control trials are assumed to be perceived as sound from the center; in fact, they elicit few head turns, as do precedence-effect trials. Although single-source trials from the side have never failed to produce statistically reliable head turns in our studies, the proportion of single-source trials on which head turning was observed decreased as the overall probability of lateral sound trials decreased. That is, having fewer opportunities during the testing session to make a head turn toward sound seemed to suppress head turning on those trials when it was expected.

To investigate this observation systematically, we varied the proportion of lateral to midline sound presentations from .25 to .75 (Clarkson, Morrongiello, & Clifton, 1982). We found that newborns showed a linear increase in the proportion of head turns on lateral trials as these trials occupied proportionately more of the session, whereas head turning on midline trials was unaffected by trial-type probability. Specifically, head turning occurred on 12.5%, 29.2%, and 41% of lateral trials with presentation probabilities of .25, .50, and .75, respectively; for midline trials, head turning averaged only 5% and was unrelated to presentation probabilities. These findings imply that reliable head turning toward lateral sound will not be recruited unless fairly high proportions (50% or greater) of lateralized stimuli are employed. If spontaneous head turning toward sound were to be developed as an audiological tool, investigators should bear in mind the interdependence of lateral sound probability and performance of the head turn response. (For further discussion of this point and others relevant to audiological testing, see Muir & Clifton, in press.)

Hypothesis 2: Temporal Parameters Influencing the Precedence Effect Will Differ among Infants, Preschoolers, and Adults

The studies supporting Hypothesis 1 sought to establish developmental changes between birth and 6 months in localizing the leading sound source. In order to describe and understand the infants' responses to precedence-effect stimuli, we needed fuller data from human adults. Following the classic contributions of Wallach *et al.* (1949) and Haas (1979), fewer than a dozen articles appear to have been published on this phenomenon with adults, and only one with children. As the infant studies represented our own initial research in this area, we felt that naive adults should be assessed under the same stimulus conditions and laboratory setting as infants, rather than relying on our own perceptions as the standard adult response.

In addition, research on the precedence effect beyond early infancy seemed likely to yield developmental differences. Other perceptual phenomena including illusions continue to change throughout childhood (Coren & Girgus, 1978). Auditory fusion (Davis & McCrosky, 1980) and perception of temporal order (Jusczyk, Pisoni, Walley, & Murray, 1980) have been found to undergo developmental changes. Davis and McCrosky (1980) reported that auditory-fusion thresholds decreased between 3 and 9 years of age; that is, older children could detect two pulses separated by shorter interpulse intervals compared to younger children. Three-year-olds required interpulse intervals of 22 to 24 msec in order to perceive two pulses, whereas the minimum interval for 9-year-olds was around 6 to 8 msec. Although auditory fusion does not involve discrimination of binaural time differences, these data aid in making predictions for the precedence effect. We hypothesized that younger subjects would require longer delays between stimulus onsets in order to perceive the lagging sounds.

Six-month-olds, 5-year-olds, and adults were compared on their responses to delay between signal onsets, a temporal parameter central to perception of the precedence effect (Morrongiello, Kulig, & Clifton, 1984). In the infant work reported thus far in this chapter, the delay between signal onsets was selected from the range of delays that reliably and unequivocally produces the precedence effect for adults. Recall that the time difference between onsets can vary widely (about 4 to 40+ msec), with the illusion of one sound at one location still maintained. Beyond the upper boundary, one localizes the lagging sound at its source, but perceives it faintly at first, then louder as the delay is increased beyond threshold level. Brief stimuli (e.g., clicks) lasting a few milliseconds appear to have a more narrow range for producing the precedence effect, whereas more complex stimuli, such as speech and music, extend the range to 40 msec or more (Wallach *et al.*, 1949). In the present study we manipulated delays so as to establish the threshold for hearing the lagging sound at each age. For 5-year-olds and adults we also varied complexity of the signal, comparing threshold delays for a click and a rattle sound. Both stimuli had been previously used in infant studies and are described under "General Methodology."

Adults (N = 48) and 5-year-olds (N = 32) received an ascending and descending trial series (counterbalanced for order across subjects for each age level), followed by a method of constant stimuli series. In the ascending series, the time delay between signal onsets was increased in 2-msec steps, starting at 2 msec and ending with 24 msec. For the rattle stimulus, the ascending series was extended to 50 msec with increases in 5-msec steps between 25 and 50 msec. The descending series was the reverse of the ascending series. All subjects received the full range of both series, including four "catch" trials in each series that contained sound from only one loudspeaker.

The method of constant stimuli procedure immediately followed, and the range of stimuli employed was determined by performance on the ascending and descending series. Each subject's threshold range was determined by (1) the delay interval at the start of three consecutive trials in the ascending series in which the subject reported hearing both loudspeakers, and (2) the delay interval at the start of three consecutive trials in the descending series when only one loudspeaker was reported. The method of constant stimuli series entailed the random presentation of delays within this threshold range in 1- and 2-msec divisions for the click and rattle stimuli, respectively. Half of each age group received the click stimulus and half received the rattle. All testing was carried out in a sound-deadened, sound-attenuated chamber.

Adults and children were told that they would hear a 10-sec train of clicks (or 10 sec of rattle sound) from one or both loudspeakers. Their task was to decide where the sound was coming from, either one or both directions. They were explicitly told that sound from one direction might sound louder than from the other direction, and to report all sounds regardless of relative loudness. (On all trials stimuli were delivered at equal intensity to both loudspeakers.) Intertrial intervals were 8 to 10 sec from the offset of one trial to the onset of the next.

A mean threshold score for each subject was calculated from the method of constant stimuli series by averaging the shortest delay on which both loudspeakers were heard and the longest delay on which one loudspeaker was heard. The average threshold scores and ranges are presented in Table 1 separated into groups by age and type of stimulus. The obvious differences in means and distributions were upheld by an analysis of variance: children and adults do not

Table 1. Average Threshold Scores (in msec) on the Method of Constant Stimuli Series[a]

	Click stimulus	Rattle stimulus
Adults	12.98	25.16
	(9.5–17.5)	(21.0–29.0)
Children	12.37	30.37
	(7.0–17.0)	(27.0–33.0)

[a]The ranges of individual threshold scores are given in parentheses.

differ in threshold for hearing the lagging sound for clicks, but children have significantly higher threshold for the rattle sound. As predicted, both age groups had a higher threshold for the rattle compared to the click.

On this latter point, no clear interpretation can be given because of the numerous differences in the two stimuli. Our study was not designed to investigate the issue of signal complexity as it relates to threshold. The click and rattle stimuli were chosen for the more practical purpose of comparing the infant data with children's and adults' performance on the range of delays that would produce the precedence effect in a particular experimental situation. However, the problem demands investigation in its own right, with manipulation of onset time, stimulus duration, and spectral cues as candidates for determinants of threshold for hearing the lagging sound. That the more complex stimulus allowed the expression of a developmental difference between preschoolers and adults, whereas the click did not, suggests that further research on threshold determinants might shed light on this developmental difference. Speech sounds, another complex signal, may be perceived more accurately by adults than young children, with some evidence that specific production capabilities are related to children's perceptual difficulties (Strange & Broen, 1981; Tallal & Stark, 1981).

To assess infants' threshold for the lagging sound, a completely different technique from that used with older subjects was necessary. The conditioned-head-turning task pioneered by Wilson (1978) has been varied and used extensively in investigations of infants' auditory discrimination (e.g., see in this volume Eilers & Oller, Chapter 11; Schneider & Trehub, Chapter 5; Trehub, Chapter 10). Aslin and Pisoni (1980) added a staircase procedure that involves a gradual increase in difficulty of the auditory discrimination being asked of the infant.

To determine the precedence-effect threshold, the following procedure was administered to 12 infants, 6 months of age (mean = 26 weeks). A background sound of clicks (3-msec duration, rate of 2 clicks/sec) was present when the infant entered the testing chamber and continued throughout the experimental session. This background click had a 7-msec delay between signal onsets, which previous research had shown to be effective in suppression of the lagging sound at this age. Infants apparently localized the sound only at the leading loudspeaker, and rapidly habituated head turns toward this repetitive sound. Trial onsets were produced by abruptly lengthening the delay interval to 42 msec, resulting in perception of the lagging click at its true location. Infants usually turned toward this "new" sound spontaneously, but if not, their behavior was shaped by an experimenter sitting opposite them who ceased entertaining them with toys and looked and pointed toward the lagging loudspeaker. If infants turned correctly, a visual reinforcer (mechanical monkey or colored-light display) was activated for a few seconds. Leading and lagging loudspeakers were located 90° off midline to the infants' right and left. Location was counterbalanced across subjects for position of leading and lagging sounds.

This shaping phase served to teach the infant to turn toward the source whenever clicks from the lagging loudspeaker were perceived. Trials were terminated by shortening the delay back to 7 msec, with these clicks continuing during the intertrial interval. The criterion for this phase was two successive head turns toward the lagging click without any cues from the experimenter; that is, she kept up the toy entertainment, not changing her behavior in any way during a trial.

The staircasing procedure or test phase immediately followed the infant's attaining criterion. During this phase the experimenter and parent wore headphones through which white noise was delivered in order to mask trial presentations. Trials were grouped into blocks of four, containing two positive and two negative trials. The two positive trials had the particular delay being tested during that block, while the two negative trials simply continued the background clicks. The negative trials served both to estimate spontaneous head turning and also to provide trials on which the experimenters scored behavior without knowing whether a test trial had been presented. Selection of trial type and delivery of reinforcement for correct head turns were controlled by a North Star Horizon microprocessor. Reinforcement was delivered when the experimenter, playing with toys, and another experimenter (blind to trial type), who monitored infants' behavior on a TV screen in an outside equipment room, both pressed buttons agreeing on a head turn. During the test phase, criterion for progressing to a lower delay was correct head turns on both positive trials and no more than one head turn on the negative trials.

The test phase began with the 42-msec delay interval for one block. If the infant met the criterion, the delay between onsets was shortened in 5-msec steps until criterion was failed. At this point, the delay interval was increased by 5 msec; if criterion was met by this reversal, the delay was decreased again. Two or more successive reversals around the same delay interval terminated the session. An infant's threshold was defined as the delay interval midway between the intervals reliably succeeded and failed.

Table 2. Threshold Scores (in msec) for 6-Month-Old Infants (arranged in order of magnitude)

*S*1	19.5	*S*7	24.5
*S*2	19.5	*S*8	24.5
*S*3	19.5	*S*9	29.5
*S*4	19.5	*S*10	29.5
*S*5	24.5	*S*11	34.5
*S*6	24.5	*S*12	34.5

Table 2 lists individual thresholds for all infants. Mean threshold was 25.33 msec, with a range of 19.5 to 34.5 msec. The number of test blocks to reach threshold varied between 6 and 13, with a mean of 9.75 blocks. Negative trials elicited few false alarms (4.7% of trials), so infants had learned to respond to the sound cue rather than simply increasing their head turning activity. Figure 1 presents the infants' data along with the preschoolers' and adults' descending threshold scores, as these scores seemed more comparable to the procedure used with infants than thresholds established by either the ascending series or method of constant stimuli. (The descending threshold mean was within 1 or 2 msec of the method of constant stimuli mean, as can be seen by comparing the data of Table 1 with that of Figure 1.) The click data for children and adults were compared to the infants' data in an age (3) × sex (2) analysis of variance. Only

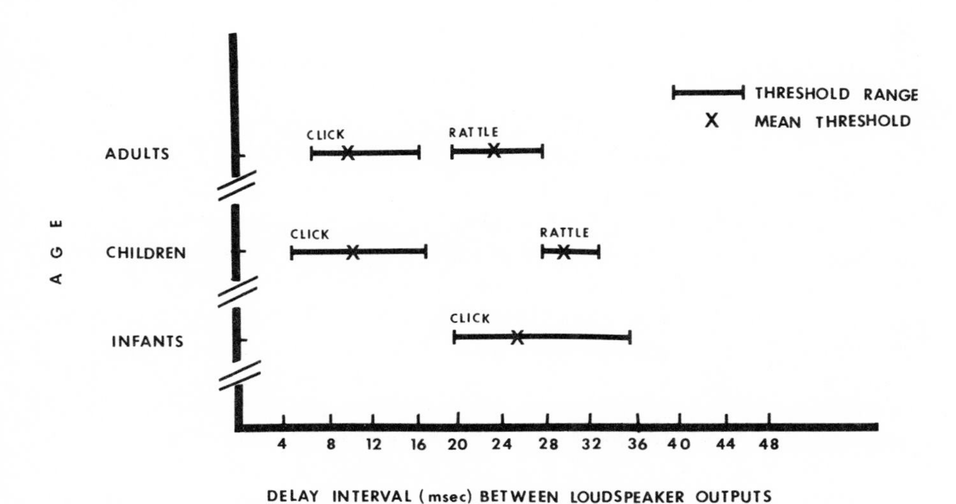

Figure 1. Comparison of infants' threshold and range with children's and adults' threshold established in the descending series. (From Morrongiello, Kulig, & Clifton, 1984.)

age was significant ($F = 70.32$, $df = 2,46$, $p < .001$), an effect readily predicted from Figure 1, which illustrates the nonoverlapping ranges between infants and older subjects.

Although these results supported the expectation that younger subjects would require longer delay intervals to perceive the lagging click, several cautions must be observed. The data from infants were necessarily derived from different procedures and cannot be truly comparable across ages. Particularly with infants, we should only conclude what they *can* hear, never what they cannot hear. Perhaps they heard a very soft click just below where threshold was determined, but this sound failed to elicit head turning. The conditioning procedure combats this argument somewhat in that a soft click might not elicit spontaneous head turning, but could elicit a conditioned head turn due to motivational properties.

Another consideration lies in infants' hearing sensitivity. Recent studies have shown that 6-month-old infants have higher thresholds for some octave-band noises compared to older children and adults (Schneider, Trehub, & Bull, 1980; Trehub, Schneider, & Endman, 1980). Although we have no comparable data on sensitivity to broad-band clicks, infants may have experienced clicks from both leading and lagging loudspeakers as less intense than older subjects did. If so, their threshold for hearing the lagging click might have been affected. The question of how overall stimulus intensity affects the precedence-effect threshold for any listener is relevant to this point; unfortunately, we know of no data on this problem.

Summary

A developmental change was predicted and found for infants' response to the precedence effect. When sound came from one spatial location, both newborns and 5-month-olds oriented toward that direction. When sound was presented through two loudspeakers with one onset delayed by 7 msec, 5-month-olds, like adults, responded to sound from the leading side only, and thus apparently perceived the precedence effect. In contrast, newborns did not turn in either direction to the more complex stimulus created by the precedence effect.

Sensory inhibition of the delayed signal produces the perception of sound localized solely on the leading side. Presumably the newborns' inability to respond in the appropriate direction in this situation reflects a failure in sensory inhibition. Because the auditory cortex has been implicated as a necessary component in this binaural suppression process, we have suggested that the observed developmental change is related to the growth in this brain area (Clifton *et al.*, 1981). Although direct attributions between brain function and particular anatomical developments in human infants cannot be made, research can be guided and predictions made concerning experimental outcomes. In addition to the precedence effect, other complex auditory tasks that require cortical processing should be investigated to determine if a similar developmental course exists over the first six months of life. Candidates for such research could be drawn from visual and cross-modal as well as auditory tasks, the criteria being those whose neurophysiological and neuroanatomical bases are well understood from animal research. Clinical studies of the same perceptual tasks further enrich hypotheses and interpretations of developmental data. This research strategy should lead to a better understanding of the ontogeny of normal auditory functioning in addition to laying the groundwork for new assessment techniques with infants and very young children.

When 6-month-old infants, preschoolers, and adults were tested on threshold for hearing a lagging click (i.e., the breakdown point for the precedence effect), infants were found to require longer delays between signal onsets than either group of older subjects. Children and adults did not differ in threshold for the click stimulus, but children required longer delays when the stimulus was a rattle sound. Both groups needed a longer delay interval for the rattle relative to the click in order to report the lagging sound.

The higher fusion point for younger children might be due to structural changes in the maturing central auditory system, or the differences in attentional and learning processes. Although the latter alternative cannot be ruled out, the nature of the precedence effect would seem to favor the former explanation. For example, the suppression of echoes occurs automatically and ubiquitously in everyday environments and does not require attention to be focused on the sound. As in other illusions, knowledge of the lagging sound in no way inhibits its suppression: one cannot avoid the illusion by trying to hear the delayed sound.

In explaining age changes in visual illusions Pollack (1969, 1970) hypothesized that changes in retinal efficiency reduce the power of certain illusions with age. Coren and Girgus (1978), in a careful discussion of these data, concluded that the magnitude of some illusions decreases with age due to structural changes (pp. 95f). Although a similar explanation of structural factors rather than attentional factors seems likely for the developmental data presented on precedence-effect thresholds, firm conclusions cannot be drawn nor can the structures involved be specified at this time. Much additional research on temporal phenomena, such as the precedence effect, auditory fusion, and temporal order, are needed to build a theoretical base before we can describe and understand auditory development.

ACKNOWLEDGMENTS

I am grateful to Barbara Morrongiello, John Kulig, John Dowd, and Marsha Clarkson for their help on all aspects of this research. I thank Gilbert Tolhurst for his advice and help in preparing auditory stimuli.

References

Aslin, R. N. & Pisoni, D. B. Some developmental processes in speech perception. In G. Yeni-Komshian, J. F. Kavanagh, & C. A. Ferguson (Eds.), *Child phonology: Perception and production.* New York: Academic Press, 1980.

Clarkson, M. G. & Berg, W. K. Cardiac deceleration and vowel discrimination in newborns: Crucial parameters of acoustic stimuli. *Child Development,* 1983, *54,* 162–171.

Clarkson, M., Morrongiello, B., & Clifton, R. Stimulus-presentation probability influences newborn head orientation to sound. *Perceptual and Motor Skills,* 1982, *55,* 1239–1246.

Clifton, R., Morrongiello, B., & Dowd, J. A developmental look at an auditory illusion: The precedence effect. *Developmental Psychobiology,* 1984, *17,* 519–536.

Clifton, R., Morrongiello, B., Kulig, J., & Dowd, J. Newborns' orientation toward sound: Possible implications for cortical development. *Child Development,* 1981, *52,* 833–838.

Coren, S. & Girgus, J. *Seeing is deceiving: The psychology of visual illusions.* Hillsdale, N.J.: Erlbaum, 1978.

Cranford, J., Ravizza, R., Diamond, I., & Whitfield, I. Unilateral ablations of the auditory cortex in the cat impairs complex sound localization. *Science,* 1971, *172,* 286–288.

Cranford, J. & Oberholtzer, M. Role of neocortex in binaural hearing in the cat. II. The "precedence effect" in sound localization. *Brain Research,* 1976, *111,* 225–239.

Davis, S. & McCrosky, R. Auditory fusion in children. *Child Development,* 1980, *51,* 75–80.

Dekaban, A. *Neurology of early childhood.* Baltimore: William & Wilkins, 1970.

Emde, R., Gaensbauer, T., & Harmon, R. Emotional expression in infancy: A biobehavioral study. *Psychological Issues,* 1976, *10,* No. 37.

Erulkar, S. Comparative aspects of spatial localization of sound. *Physiological Review,* 1972, *52,* 237–337.

Field, J., Muir, D., Pilon, R., Sinclair, M., & Dodwell, P. Infants' orientation to lateral sounds from birth to three months. *Child Development,* 1980, *51,* 295–298.

Green, D. *An introduction to hearing.* Hillsdale, N.J.: Erlbaum, 1976.

Haas, H. The influence of a single echo on the audibility of speech. Reprinted in E. Schubert (Ed.), *Benchmark papers in acoustics, Vol. 13: Psychological acoustics.* Stroudsburg, Pa.: Dowden, Hutchinson, & Ross, 1979.

Heffner, H. *The effect of auditory cortex ablation on sound localization in the monkey* (Macaca mulatta). Unpublished dissertation, Florida State University, 1973.

Hochster, M. & Kelly, J. The precedence effect and sound localization by children with temporal lobe epilepsy. *Neuropsychologia,* 1981, *19,* 49–55.

Jusczyk, P., Pisoni, D., Walley, A. & Murray, J. Discrimination of relative onset time of two-component tones by infants. *Journal of the Acoustical Society of America,* 1980, *67,* 262–270.

Kelly, J. Localization of paired sound sources in the rat: Small time differences. *Journal of the Acoustical Society of America,* 1974, *55,* 1277–1284.

Klump, R. & Eady, H. Some measurements of interaural time difference thresholds. *Journal of the Acoustical Society of America,* 1956, *28,* 859–860.

Mills, A. Auditory localization. In J. V. Tobias (Ed.), *Foundations of modern auditory theory* (Vol. 2). New York: Academic Press, 1972.

Morrongiello, B., Clifton, R., & Kulig, J. Newborn cardiac and behavioral orienting responses to sound under varying precedence effect conditions. *Infant Behavior & Development,* 1982, *5,* 249–259.

Morrongiello, B., Kulig, J., & Clifton, R. Developmental changes in auditory temporal perception. *Child Development,* 1984, *55,* 461–471.

Muir, D. & Clifton, R. K. Infants' orientation to the location of sound sources. In G. Gottlieb & N. A. Krasnegor (Eds.), *Measurement of audition and vision in the first postnatal year of life: A methodological overview.* Norwood, N.J.: Ablex, in press.

Muir, D. & Field, J. Newborn infants orient to sounds. *Child Development,* 1979, *50,* 431–436.

Muir, D., Abraham, W., Forbes, B., & Harris, L. The ontogenesis of an auditory localization response from birth to four months of age. *Canadian Journal of Psychology,* 1979, *33,* 320–333.

Pollack, R. Some implications of ontogenetic changes in perception. In J. Flavell & D. Elkind (Eds.), *Studies in cognitive development: Essays in honor of Jean Piaget.* New York: Oxford University Press, 1969.

Pollack, R. Mueller-Lyer illusion: Effect of age, lightness contrast, and hue. *Science,* 1970, *170,* 93–95.

Schneider, B., Trehub, S., & Bull, D. High-frequency sensitivity in infants. *Science,* 1980, *207,* 1003–1004.

Strange, W. & Broen, P. The relationship between perception and production of /W/, /r/, and /1/ by three-year-old children. *Journal of Experimental Child Psychology,* 1981, *31,* 81–102.

Tallal, P., & Stark, R. Speech acoustic-cue discrimination of normally developing and language-impaired children. *Journal of the Acoustical Society of America,* 1981, *69,* 568–574.

Trehub, S., Schneider, B., & Endman, M. Developmental changes in infants' sensitivity to octave-band noises. *Journal of Experimental Child Psychology,* 1980, *29,* 282–293.

Wallach, H., Newman, E., & Rosenzweig, M. The precedence effect in sound localization. *American Journal of Psychology,* 1949, *62,* 315–336.

Whitfield, I., Auditory cortical lesions and the precedence effect in a four-choice situation. *Journal of Physiology,* 1978, *289,* 81.

Whitfield, I., Cranford, J., Ravizza, R., & Diamond, I. Effects of unilateral ablation of auditory cortex in cat on complex sound localization. *Journal of Neurophysiology,* 1972, *35,* 718–731.

Whitfield, I., Diamond, I., Chiveralls, K., & Williamson, T. Some further observations on the effects of unilateral cortical ablation on sound localization in the cat. *Experimental Brain Research,* 1978, *31,* 221–234.

Wilson, W. Behavioral assessment of auditory function in infants. In F. D. Minifie & L. L. Lloyd (Eds.), *Communicative and cognitive abilities—Early behavioral assessment.* Baltimore: University Park Press, 1978.

Yakovlev, P. & Lecours, A. The myelogenetic cycles of regional maturation of the brain. In A. Minkowski (Ed.), *Regional development of the brain in early life,* Philadelphia: F. A. Davis, 1967.

CHAPTER 5

Behavioral Assessment of Basic Auditory Abilities

Bruce A. Schneider and Sandra E. Trehub

Centre for Research in Human Development
Erindale College
University of Toronto
Mississauga, Ontario

Until very recently, little was known about the development of basic auditory abilities in infancy despite the acknowledged importance of these abilities for the perception of complex auditory patterns such as speech (see Schneider, Trehub, & Bull, 1979; Trehub, Bull, & Schneider, 1981a). The research in infant audition focused primarily on the neonatal period, with little or no attention to subsequent development. This research direction was motivated largely by clinical concerns, such as devising procedures for early identification of hearing loss, and, consequently, led to the use of a variety of response measures (e.g., auropalpebral reflex, acoustic reflex, Moro reflex, nonnutritive sucking, changes in general activity, heart rate, skin resistance changes, and evoked responses). In each of the various response systems there is a function that relates response magnitude to sound intensity. These response–intensity functions, however, tend to differ and to yield, therefore, conflicting estimates of auditory sensitivity (Schneider *et al.*, 1979, Schneider & Trehub, 1984).

In our own laboratory, the focus has been on precise specification of the course of development. Accordingly, we sought a procedure that was (1) applicable over a wide age range, (2) non-invasive, and (3) sufficiently robust to provide data within a single session. A likely technique was one that involved visual reinforcement of a localization or head-turning response (Moore, Thompson, & Thompson, 1975). The localization response or orientation of head and eyes toward a sound source is readily elicited by about 4 months of age (Muir, Chapter 3) but can only be effectively conditioned at about 5 to 6 months of age (Moore *et al.*, 1975), when it becomes particularly useful as a psychophysical technique (Trehub, Schneider, & Bull, 1981).

In our version of this technique, both the parent (seated on a test chair) and the infant (placed on the parent's lap) face an experimenter seated in the opposite corner of a sound-attenuating booth. The parent and experimenter wear headphones over which masking noise is presented to prevent them from detecting the locus of a test signal. A trial is initiated only when the infant is quiet and is looking directly ahead, at which time the experimenter in the booth presses a button to initiate a trial. A signal is then presented on one of two speakers located 45 degrees to the infant's left and right. The signal remains on until the infant makes a head turn of 45 degrees or more toward either side, at which time the experimenter presses one of two buttons to indicate the direction of the head turn. The signal is then turned off and, if the head turn is in the direction of the speaker producing the signal, a toy above that speaker is illuminated and activated for a period of 4 sec. If the head turn is in a direction away from that speaker, there is a 4-sec silent interval.

To ensure that all the infants can perform the response of turning to the sound location, a training criterion is employed with sound intensity well above threshold. During the training period, the location of the signal is alternated between left and right speakers until the child makes four successive correct responses. The sound intensity is then reduced 5 or 10 dB, and the alternation continued until the infant again makes four successive correct responses. When the training criterion is reached, the actual test series begins. Typically, more than 95% of the infants between 6 and 18 months can satisfy this criterion and 85% to 90% complete a session without fussing or crying. During a test session, four or five different levels of the signal are presented (levels and location randomized) a total of 5 times each. At the conclusion of the test session, we often attempt a second session in order to maximize the amount of information obtained from any single infant. At the beginning of the second session, a training criterion is again introduced with the exception that only 2 correct responses at each level are required.

Measuring Absolute Thresholds

To date, we have used this technique to determine absolute thresholds (Schneider, Trehub, & Bull, 1980; Trehub, Schneider, & Endman, 1980), masked thresholds (Bull, Schneider, & Trehub, 1981; Trehub, Bull, &

Schneider, 1981b), and incremental thresholds (Schneider, Bull, & Trehub, in preparation) in infants from 6 to 24 months of age. In the first of these studies (Trehub *et al.*, 1980) thresholds were determined for octave-band noises with center frequencies of 200, 400, 1,000, 2,000, 4,000, and 10,000 Hz for infants 6, 12, and 18 months of age. The upper panel of Figure 1 shows the percentage of correct head turns as a function of the decibel level of the octave-band noises for the six different test frequencies, with each point being based on a minimum of 85 trials. (In all subsequent figures, each point is based on at least 100 trials.) It is clear that as the intensity level increases, the percentage of correct responses also increases but never quite reaches 100%, even at the higher levels. This may be attributed to momentary lapses of attention. The bottom panel of Figure 1 presents psychometric functions for two adults who were tested in a similar manner at frequencies of 400, 1,000, 4,000, and 10,000 Hz. Since each point of adult data is based on only 40 trials, these adult functions are more variable at the lower intensity levels than the infant functions.

Figure 2 indicates threshold values (defined as 65% correct) as a function of frequency for all age groups. It can be seen that the threshold function for 12- and 18-month-old infants is similar across the frequency range studied, but the 6-month-old group is approximately 5 to 8 dB less sensitive than the older infants at the lower frequencies. At the higher frequencies, there is no substantial difference between groups. Octave-band threshold values can also be seen for the

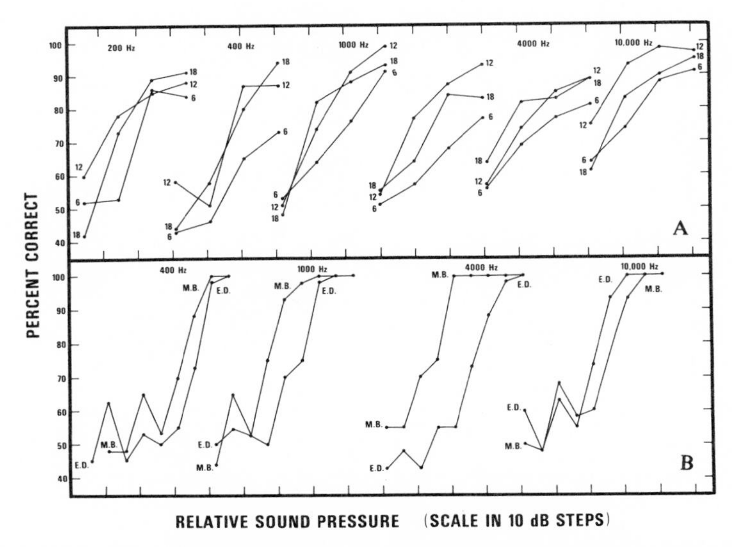

Figure 1. (A): Percentage of correct head turns as a function of decibel level of 6 test frequencies for infants 6, 12, and 18 months. (B): Percentage of correct responses as a function of decibel level of 4 test frequencies for two adults. Stimuli are octave-band noises. From Trehub *et al.* (1980).

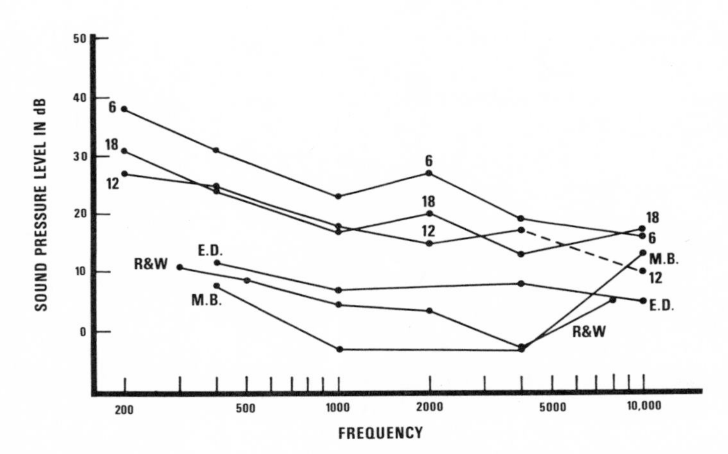

Figure 2. Threshold as a function of frequency for infants 6, 12, and 18 months of age and for two adults. Thresholds determined by Robinson and Whittle (1964) are also plotted. From Trehub *et al.* (1980).

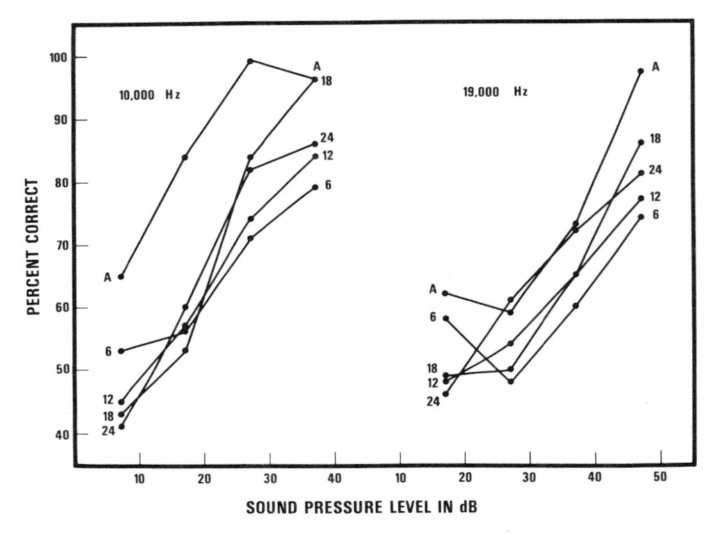

Figure 3. Percentage of correct head turns as a function of decibel level of half-octave band noises with center frequencies of 10,000 and 19,000 Hz for infants 6, 12, 18, and 24 months of age and for adults. From Schneider *et al.* (1980).

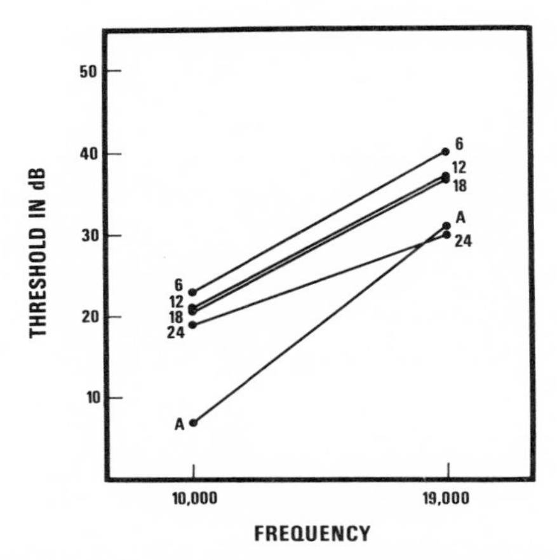

Figure 4. Thresholds for half-octave band noises with center frequencies of 10,000 and 19,000 Hz for infants 6, 12, 18, and 24 months of age and for adults. From Schneider *et al.* (1980).

two adult subjects, as well as for other adults tested by Robinson and Whittle (1964) in frontal-incidence conditions using a method of limits. The threshold functions for adult subjects are much flatter than for infants. At the lower frequencies, differences between infant and adult thresholds can exceed 20 to 25 dB; at higher frequencies, however, thresholds for the adults and infants begin to converge.

These findings encouraged us to extend our work to even higher frequencies (Schneider, Trehub, & Bull, 1980). We tested infants (6–24 months of age) and adults with each of two half-octave bands of noise centered at 10,000 and 19,000 Hz. As can be seen in Figure 3 the psychometric functions for the four infant groups at 10,000 Hz are quite similar and are approximately 15 dB higher than the adult function. At 19,000 Hz, however, this adult–infant difference is substantially attenuated and, in fact, thresholds of the 24-month-old infants appear to be equivalent to those of adults (see Figure 4).

These two studies indicate that the disparity between infant and adult thresholds decreases as the frequency of the noise band increases, culminating in comparable thresholds at 19,000 Hz. Thus, there is suggestive evidence that developmental changes in auditory sensitivity beyond 6 months of age are reflected largely in improvement at the lower frequencies.

Masking in Infants

It is interesting to note that masking, or the detection of signals in noise, can also be studied by using our two-alternative, forced-choice technique. Masking is of interest because of its relevance to natural auditory phenomena and its potential for elucidating the analytical properties of the developing ear. On the basis of recent noise surveys, it has been suggested that current levels of environmental noise may interfere with speech, language, and listening skills in children (Mills, 1975). Thus, on the one hand, it is important from a practical viewpoint to ascertain the levels of background noise that are likely to impede the development of speech and language. On the other hand, masking studies can be used to elucidate the properties of the developing ear and, in particular, to reject some possible explanations of adult–infant threshold differences that we have obtained.

To study masking, our procedure is modified simply by having both speakers carry a broadband masker continuously. In Bull *et al.* (1981), the signal was an octave-band noise with a center frequency of 4,000 Hz. Infants 6, 12, 18, and 24 months of age were tested at two different levels of background noise, 42 and 60 dBC, as were adults. Figure 5 shows the percentage of correct responses as a function of the decibel level of the octave-band signal for the two background noise levels for the five age groups. The psychometric functions are somewhat similar except that the infant functions are shifted 10 to 20 dB to the right of the adult functions.

Figure 6 shows the threshold values for the two masking conditions at the five age levels. Thresholds are similar for the 12-, 18-, and 24-month-old infants at both levels of background noise, but are approximately 7 to 8 dB higher for the 6-month-old group. The infant groups are considerably less sensitive (16–25 dB) than the adults. For all age levels, however, the effect of increasing the masker

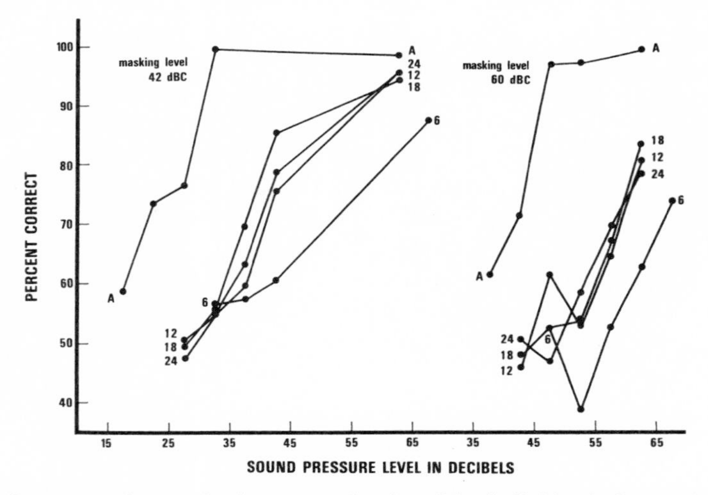

Figure 5. Percentage of correct head turns as a function of the decibel level of octave-band noise under two masking conditions for infants 6, 12, 18, and 24 months of age, and adults. From Bull *et al.* (1981).

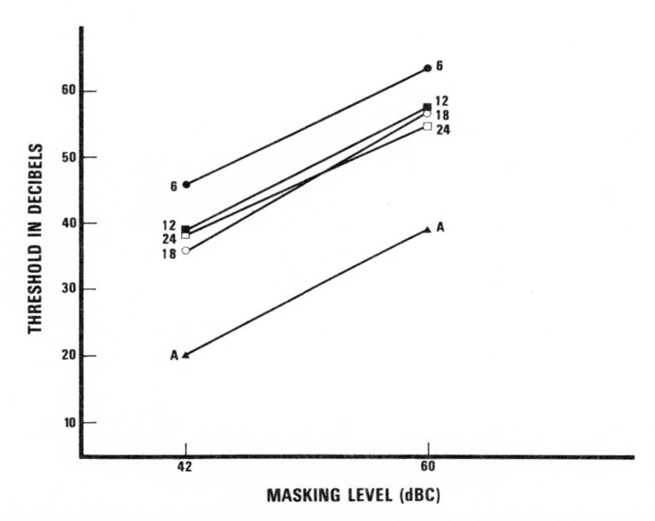

Figure 6. Thresholds as a function of masking level for infants 6, 12, 18, and 24 months of age, and adults. From Bull *et al.* (1981).

by 18 dB is to raise the thresholds approximately 18 dB (an average of 18.6 dB for the infants and 19 dB for the adults). This result is consistent with Hawkins and Stevens' (1950) classic finding that an x dB increase in masked threshold occurs when the masking noise is increased by x dB. Thus, infants and adults respond similarly to changes in the level of the masker.

Thresholds for infants were from 16 to 25 dB higher than those for adults at all masking levels, which is comparable to the adult–infant difference in the *absence* of masking noise (Trehub *et al.*, 1980). Thus, there appears to be adult–infant differences that are *independent* of the level of background noise.

Proponents of neural models of the detection of signals in noise (e.g., Zwislocki, 1964) have argued that detection depends on the *internal* signal-to-noise ratio; that is, a signal is detected when the ratio of the neural effect of the signal to the neural effect of the masker is constant. Since the detection of a signal in quiet is not perfect, it is necessary to assume the existence of some intrinsic background noise. Thus, the threshold for a signal in quiet reflects the detection of this signal against this intrinsic background noise and presumably occurs when the ratio of the neural effect of the signal to the intrinsic background noise reaches the specified or criterial signal-to-noise ratio. In general then, the signal is detected when

$$f(I_S)/[g(I_N) + B] = c \qquad (1)$$

where $f(I_S)$ is the internal effect of the signal (I_S), $g(I_N)$ is the internal effect of the noise (I_N), and B is the amount of background neural activity in quiet. If we assume that f and g are both identical power functions of sound intensity, we would have

$$I_S^n \,/\, (I_N^n + B) = c \tag{2}$$

where n is a constant. Note that when the background level is sufficiently high, detection will occur when the signal-to-noise ratio in the physical domain is approximately constant. For in that instance,

$$(I_S^n \,/\, I_N^n) \cong c \tag{3}$$

This model is, of course, in accord with the Hawkins and Stevens' finding that an x dB increase in masking noise results in an x dB increase in masked threshold.

It is interesting to note that, according to this model, the adult–infant differences observed in a quiet background could result from two sources. Infants could have a higher background noise level than adults (a larger B); the internal signal-to-noise ratio required for threshold could be greater in infants (a larger c); or both B and c could be larger for infants. The infant-masking data argue strongly against the first possibility. If the adult–infant threshold difference in quiet were due simply to a higher background noise, then equation 3 would indicate that the adult–infant difference would disappear when high levels of masking noise were employed. But there is no suggestion in Figure 6 that the adult–infant differences are starting to disappear at the higher masking levels.

It is also interesting to note that the data are *not* consistent with what we would expect if infants were simply subject to a conductive hearing loss. Suppose infants experienced a 20-dB conductive hearing loss compared to adults. In a quiet background (noise intensity equal to zero), equation 2 would indicate that the signal intensity for infants would have to be 20 dB higher than for adults in order to arrive at the same value of c. However, for the detection of signals in very noisy backgrounds, where equation 3 holds, there should be no adult–infant difference. Equation 3 shows that at high levels of masking noise, threshold is reached when the physical signal-to-noise ratio is constant. A conductive hearing loss would attenuate both signal and noise intensities leaving the effective signal-to-noise ratio unchanged. If adult–infant differences at threshold are to be attributed solely to a conductive hearing loss, then these threshold differences should disappear at high levels of background noise. Figure 6 shows that they do not, thereby rejecting the model of conductive hearing loss.

Incremental Thresholds and Binaural Phenomena

To determine incremental thresholds in infants and adults (Schneider, Bull, & Trehub, in preparation), we adapted our two-alternative, forced-choice procedure in the following manner. As in the masking situation, a constant background noise (0–5000-Hz bandwidth) was presented over the two loudspeakers. In the intensity increment experiment, however, the signal added to one of the speakers was a noise having the same power spectrum as that of the background noise. Adding a noise with the same power spectrum to a background noise is the same as incrementing that noise. Thus, the detection of broad-spectrum noise in background noise that has the same power spectrum can be analyzed as either a masking experiment or an increment detection experiment (Miller, 1947). When analyzed as a masking experiment, we simply plot the function relating correct head turns to the intensity of the signal, and determine the threshold for that signal. To convert this to an intensity–increment threshold, it is only necessary to compute how much (in dB) the signal increments the background noise at threshold. In the present experiment, two levels of background noise (37.5 and 57.5 dBC) were employed. In addition, the nature of the noise signal added to the background was varied. In one experiment, the added noise was taken from the same noise generator as the background noise; in a second experiment, an uncorrelated noise (noise from an independent signal generator) was added to one of the two speakers to increment its intensity. The nature of the noise signal (correlated vs. uncorrelated) was varied since the literature on masking level differences (e.g., Colburn & Durlach, 1978; Durlach & Colburn, 1978) suggests that thresholds for these two kinds of noise signals might differ significantly.

It is important to note that when an independent noise source is added to a background noise, the intensity of the signal plus noise combination is equal to the intensity of the noise plus the intensity of the signal, that is,

$$I_{SN} = I_S + I_N$$

where I_S, I_N, and I_{SN}, are the intensities of the signal, background noise, and signal plus noise, respectively. When a coherent noise source is added to a background noise, however, the total intensity of the signal plus noise is

$$I_{SN} = I_S + I_N + 2(I_S I_N)^{1/2}$$

Thus, adding a coherent signal of x dB to a background noise produces a different increment of sound intensity than adding an independent signal of x dB. To facilitate comparisons between Experiments 1 and 2, we computed intensity values of uncorrelated noise that would produce the same intensity increment as those produced by the correlated noise in Experiment 1. We will refer to these computed values as the equivalent independent noise values. They are the decibel values of independent noise that produce the same intensity increment as that produced by the actual values of correlated noise used in Experiment 1. These equivalent independent noise values are plotted in Figure 7.

Figure 7 shows the percentage of correct head turns as a function of the

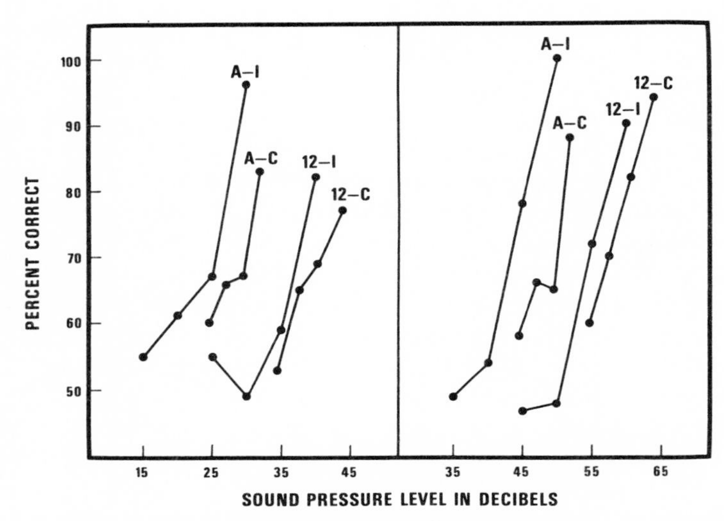

Figure 7. Percentage of correct localization responses as a function of the decibel level of the noise increment for background intensities of 37.5 dB (left) and 57.5 dB (right). A-I is adults, independent noise increment; A-C is adults, correlated noise increment; 12-I is 12-month infants, independent noise increment; 12-C is 12-month infants, correlated noise increment. From Schneider *et al.* (in preparation).

decibel level of the noise signal. Note that adults are more sensitive than infants at both levels of background noise. Furthermore, note that detection for an independent increment is better than for a correlated increment for both infants and adults; that is, the intensity increment needed for detection of the signal is less for independent increments than it is for correlated increments. This result is consistent with research on adult masking level differences; when an identical masking noise is presented to both ears, the threshold for detection of a signal presented monaurally is significantly lower than for the same signal presented binaurally. When the signal is presented monaurally, there are waveform differences in the pattern of stimulation to the two ears (one ear receives the signal plus noise, while the other ear is presented with the noise alone). When the signal is presented binaurally, however, there are no differences in waveform during signal presentation since both ears are presented with signal plus noise. Thus, the improved detection with monaural presentation can be attributed to the ability of the auditory system to respond to some aspect of waveform differences in the two ears. In the present experiment, we would expect a waveform difference when an independent noise increment is presented to one speaker during signal presentations. In that case, one loudspeaker continues to produce the background noise while the second loudspeaker produces a waveform consisting of the sum of the background noise and the independent noise. This is to be contrasted to Experiment 1 where the same noise was added to one of the loudspeakers. Here the waveforms produced by the two speakers differ only in intensity. If the listener can utilize the waveform differences that exist in Experiment 2, we would expect lower thresholds in this experiment. This is exactly what we found for both infants and adults, although the effect was somewhat stronger for adults (see Figure 7). Thus, by one year of age, infants are clearly sensitive to binaural differences in waveform, but the developmental course for binaural sensitivity to waveform differences remains to be determined.

Figure 7 can also be used to estimate the size of the intensity increment needed by infants at the two levels of background noise. For the correlated noise source, infants can detect 75% of the time a 4.76-dB increment in noise against a 40-dB background noise, whereas a 2.45-dB increment is all that is needed in a background of 60-dB noise. This demonstrates a sensitivity to noise increments that is much greater than that previously reported (e.g., Bartoshuk, 1964).

Toward a Theory of Auditory Development

To date, our behavioral evaluation of auditory abilities in infants and young children has provided us with much information on the course of development of the auditory system. First, we first found that sensitivity differences between infants and adults are greater at lower compared to higher frequencies. At very high frequencies (19,000 Hz), thresholds for infants and adults were found to be comparable. Thus, major developmental changes in sensitivity are likely to occur in the low-frequency region. Furthermore, the time-course of these changes may well be lengthy since infants of 18 months still have thresholds that are 15 to 20 dB higher than adults for the low-frequency signals. Second, our masking data suggest that infant hearing is impaired relative to adults. Since this impairment is still present at 24 months, it also suggests a long time-course for auditory development. Third, a comparison of thresholds in noise and in quiet allows us to reject two models of adult–infant hearing differences: namely, that they are due solely to differences in internal physiological noise, or that they represent a conductive hearing loss. Finally, 12-month-old infants show relatively good sensitivity to intensity increments and at least rudimentary sensitivity to higher order characteristics of the stimuli.

Two general features of these results merit further comment. First, infants typically require greater signal intensity than adults for detection in both quiet and in noise. Second, the extent of these adult–infant intensity differences may be frequency dependent. Schneider *et al.* (1980) and Trehub *et al.* (1980) found that adults were about 20 dB more sensitive than infants at the lower frequencies (less than 4,000 Hz) and that these differences disappeared at 19,000 Hz. For thresholds in quiet, then, there are adult–infant sensitivity differences that are frequency dependent. Both Bull *et al.* (1981) and Trehub *et al.* (1981b) found adult–infant differences in sensitivity for signals in noise. Unfortunately, we do not know whether these differences are frequency dependent and, if so, whether they show the same kind of frequency dependency as found for thresholds in quiet. Finally, our intensity discrimination data reveal adult–infant sensitivity differences. Again, we do not know whether these differences are frequency dependent.

Let us now consider the kind of model that can account for these data. Suppose we assume that the interaction between frequency and adult–infant sensitivity differences is independent of background noise level. If this assumption is correct, then we can account for these developmental changes by assuming that the size of the critical band is changing in a systematic fashion with age.

Several experiments have shown that when adults attempt to detect a pure tone signal in broadband noise, only the energy in a very narrow band of frequencies surrounding the pure tone contributes to its masking. It is as if the ear has a narrow band filter whose center frequency is set at the frequency of the pure tone. If the filter has a very narrow bandwidth, then the energy in the broadband noise that will contribute to the masking of the pure tone is quite small. If, instead, the filter has a wide bandwidth, then a greater proportion of the energy in the masker will contribute to the masking of the pure tone at the center of the bandwidth. Thus, the effective signal-to-noise ratio for a pure tone in a broadband masker will depend on the bandwidth of the filter. As the bandwidth increases, the signal-to-noise ratio will decrease.

If we assume that the effective signal-to-noise ratio required for detection of threshold is the same for infants and adults, but that infants have larger critical bands (broader filters) than adults, we would predict that masked thresholds for infants would be higher than those for adults. This is, of course, what we have found. It can be shown that this model would also predict adult–infant sensitivity differences for the detection of signals in a quiet background. Recall that proponents of detection models generally assume that detection depends on the *internal* signal-to-noise ratio and postulate the existence of some intrinsic background noise (see equation 1). This intrinsic background noise would presumably be spread along the cochlear partition and thus would have a broad spectrum. Accordingly, if infants have larger critical bands than adults, their effective intrinsic background noise would be greater than that of adults, and this would account for adult–infant sensitivity differences in quiet.

This model requires that adult–infant sensitivity differences be independent of the level of background noise and that the intrinsic background noise also has a relatively broad and flat spectrum. An evaluation of these assumptions requires masking functions at several frequencies and levels of background noise.

If we assume, further, that the rate at which the infant's critical band approaches the adult value is dependent on frequency, then it is possible to explain why there are large adult–infant sensitivity differences at low but not at high frequencies. Suppose, for example, that the critical bands for the higher frequencies approach adult values at an earlier age, whereas the critical bands for lower frequencies take longer to reach adult status. We would then expect substantial adult–infant threshold differences at the lower frequencies, and very small differences at the higher frequencies, the pattern that has emerged in our research.

There is some anatomical and physiological evidence that is consistent with this model of critical band development. First, the hair cells and associated neural structures mature earliest at the base of the cochlea (see Bredberg, Chapter 1). If maturation at more central levels follows the time course found for cochlear development, then we might expect that structures that process information transduced at the base of the cochlea would reach adult status earlier than structures associated with the apex. Since the base transduces high frequencies, critical bands for high frequencies would be expected to reach adult status earlier than those for low frequencies. Second, there is evidence for differential rates of maturation for turning curves of individual fibers in the cat. Romand (1979) found that between 20 and 30 days of age there was little improvement in the degree of tuning for high-frequency fibers (5%), which already approached adult values. This contrasts with the over 200% improvement in the degree of tuning for low-frequency fibers.

This critical band model of auditory development provides a speculative but economical description of the course of auditory development in infancy. Empirical evaluation of the model is our next step.

References

Bartoshuk, A. K. Human neonatal cardiac responses to sound: A power function. *Psychonomic Science,* 1964, *1,* 151–152.

Bull, D., Schneider, B. A., & Trehub, S. E. The masking of octave-band noise by broad-spectrum noise: A comparison of infant and adult thresholds. *Perception & Psychophysics,* 1981, *30,* 101–106.

Colburn, H. S., & Durlach, N. I. Models of binaural interaction. In E. C. Carterrette & M. P. Friedman (Eds.), *Handbook of perception: Hearing* (Vol. 4). New York: Academic Press, 1978.

Durlach, N. I., & Colburn, H. S. Binaural phenomena. In E. C. Carterrette & M. P. Friedman (Eds.), *Handbook of perception: Hearing* (Vol. 4). New York: Academic Press, 1978.

Hawkins, J. E., & Stevens, S. S. The masking of pure tones and of speech by white noise. *Journal of the Acoustical Society of America,* 1950, *22,* 6–13.

Miller, G. A. Sensitivity to changes in the intensity of white noise and its relation to masking and loudness. *Journal of the Acoustical Society of America,* 1947, *19,* 609–619.

Mills, J. H. Noise and children: A review of the literature. *Journal of the Acoustical Society of America,* 1975, *58,* 767–779.

Moore, J. M., Thompson, G., & Thompson M. Auditory localization of infants as a function of reinforcement conditions. *Journal of Speech and Hearing Research,* 1975, *40,* 29–34.

Robinson, D. W., & Whittle, L. S. The loudness of octave-bands of noise. *Acustica,* 1964, *14,* 24–35.

Romand, R. Development of auditory nerve activity in kittens. *Brain Research,* 1979, *173,* 554–556.

Schneider, B. A., Bull, D., & Trehub, S. E. *The detection of independent versus correlated noise increments: A comparison of infant and adult thresholds.* In preparation.

Schneider, B. A., & Trehub, S. E. Infant auditory psychophysics: An overview. In N. A. Krasnegor & G. Gottlieb (Eds.), *Measurement of audition and vision during the first year of life: A methodological overview.* Norwood, N.J.: Ablex, 1984.

Schneider, B. A., Trehub, S. E., & Bull, D. The development of basic auditory processes in infants. *Canadian Journal of Psychology,* 1979, *33,* 306–319.

Schneider, B. A., Trehub, S. E., & Bull, D. High-frequency sensitivity in infants, *Science,* 1980, *207,* 1003–1004.

Trehub, S. E., Bull, D., & Schneider, B. A. Infant speech and non-speech perception: A review and reevaluation. In R. L. Schiefelbusch & D. Bricker (Eds.), *Early language: Acquisition and intervention.* Baltimore: University Park Press, 1981.(a)

Trehub, S. E., Bull, D., & Schneider, B. Infants' detection of speech in noise. *Journal of Speech and Hearing Research,* 1981, *24,* 202–206.(b)

Trehub, S. E. & Schneider, B. A. Recent advances in the behavioral study of infant audition. In S. E. Gerber & G. T. Mencher (Eds.), *Development of auditory behavior.* New York: Grune & Stratton, 1983.

Trehub, S. E., Schneider, B. A., & Bull, D. Effect of reinforcement on infants' performance in an auditory detection task. *Developmental Psychology,* 1981, *17,* 872–877.

Trehub, S. E., Schneider, B. A., & Endman, M. Developmental changes in infants' sensitivity to octave-band noises. *Journal of Experimental Child Psychology,* 1980, *29,* 282–293.

Zwislocki, J. J. Masking: Experimental and theoretical aspects of simultaneous, forward, backward, and central masking. In E. C. Carterette & M. P. Friedman (Eds.), *Handbook of perception: Hearing.* Vol. 4. New York: Academic Press, 1978.

CHAPTER 6

Physiological Measures of Auditory Sensitivity

Near-Threshold Intensity Effects

W. Keith Berg

Department of Psychology
University of Florida
Gainesville, Florida

The scientist interested in evaluating the auditory sensitivity of young infants is presented with a difficult choice. On the one hand, physiological measures appear very attractive. There are a wide variety of them to choose from; they often can be recorded while the infant is asleep as well as awake; and the response systems are typically more mature than most externally observable motor systems. This latter quality is important because it allows the assessment of sensory-system development to be at least partially isolated from response-system development. On the other hand, physiological measures are often viewed as categorically different from behavioral measures—especially the traditional ones used with older children and adults. These effects have led Trehub and colleagues (Schneider, Trehub, & Bull, 1979; Trehub, Schneider, & Bull, 1981), and others before them, to conclude that psychophysiological techniques are sensitive to the ''significance'' of the stimulus and therefore track what they refer to as the ''attentional threshold'' rather than the threshold of audibility.

This chapter will address this viewpoint, agreeing with some aspects of it and taking issue with others. Specifically, it will be suggested that

1. Physiological measures of auditory responsivity in the intact human

The writing and research reported was supported by federal contract NINCDS NO1-NS-2131.

infant and adult may be far more sensitive than has generally been perceived.

2. Physiological measures can be very sensitive to the "significance" of the stimulus, just as Schneider *et al.* (1979) note, and this can be equally true of behavioral measures. However, with regard to measuring auditory sensitivity in the infant, this may be an asset rather than a liability.
3. Physiological measures should not be viewed in a category separate from behavioral measures, except with regard to the necessary instrumentation. Behavioral and physiological measures both have the potential to be good or poor indicators of auditory sensitivity depending on the age of the test subject, the stimulus presented, and the paradigm employed.

These points will be illustrated in an examination of some recent data on the unexpected effectiveness of low-intensity auditory stimuli in eliciting various physiological responses.

Perhaps the most typical function relating stimulus intensity and any of a variety of physiological measures (and behavioral measures as well) is one of a monotonic increase in response with stimulus intensity. An example of such a finding is shown in the right panel of Figure 1, taken from the work of Jackson (1974). What is shown is the change in skin conductance of adult listeners in response to a 1,000-Hz, 2-sec pure tone. It is evident that there is little or no response until the stimulus reaches 60 dB SPL. This measure, as used here,

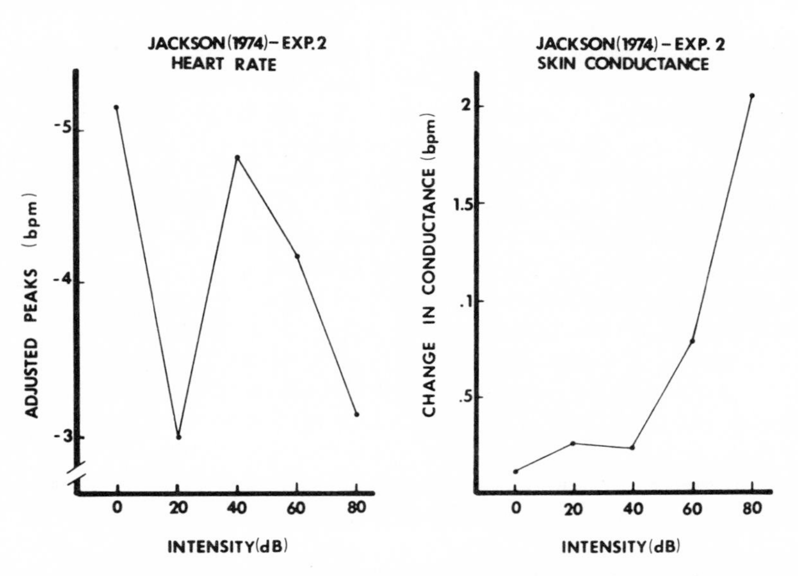

Figure 1. Adapted from experiment 2 of Jackson (1974). Note the contrast between the non-monotonic function for heart rate and the monotonic function for skin conductance.

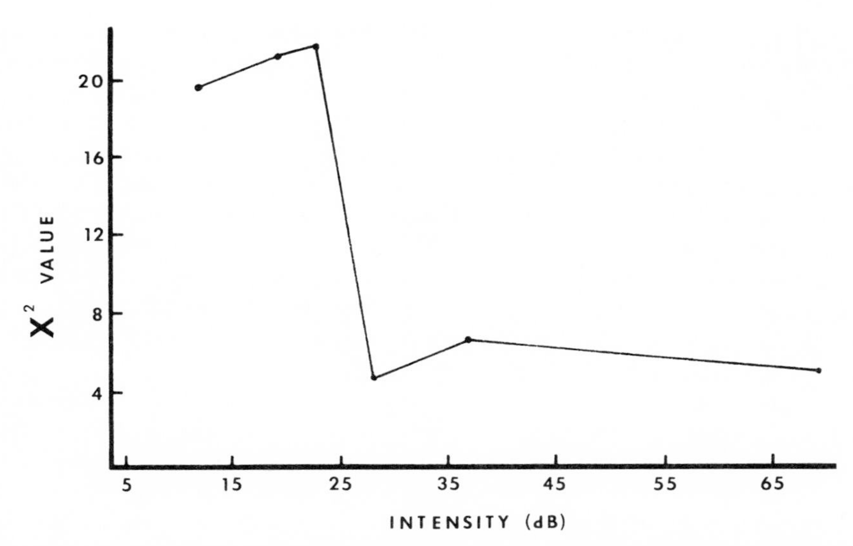

Figure 2. Adapted from Pool *et al.* (1966). The chi-square values indicate the statistical reliability of the respiratory slowing response at each stimulus intensity.

would not be very satisfactory for assessing the lower limit of auditory detectability.

This conclusion might not hold for other response systems. The concepts of Sokolov (e.g., 1963) suggest that stimuli near the auditory threshold should have a special significance and therefore elicit an enhanced response in systems that reflect the orienting reflex. One of the earliest demonstrations of the potential effectiveness of near-threshold stimuli came from the work of Rousey and colleagues. In a number of studies (e.g., Pool, Goetzinger, & Rousey, 1966; Rousey & Reitz, 1967; Rousey, Snyder, & Rousey, 1964), these investigators found that auditory stimuli within about 10 dB of behaviorally determined thresholds produced slowing of respiratory rate. Figure 2 shows data from adults adapted from the Pool *et al.* (1966) study. Although they did not present specific data on the magnitude of the slowing at each intensity, these data on the reliability of the response at each intensity (as evaluated by the chi-square statistic) clearly indicate that the effect is most consistent at very low intensities.

Although the scoring procedures and replicability of the respiratory data have been questioned (e.g., Hartley & Hetrick, 1973), the enhancement of responses to near-threshold stimuli is not limited to respiration. Graham and Clifton (1966) argued that heart-rate deceleration is the cardiac component of the orienting response. As such, it should be a good measure on which to test the hypothesis of near-threshold enhancement of orienting. Jackson (1974) provided the appropriate data by comparing heart-rate decelerations with skin-conductance changes in adult subjects. Figure 1 shows Jackson's heart-rate results alongside the skin-conductance data described earlier. These two response measures were recorded simultaneously from the same subjects. The heart-rate score is the peak deceleration in the first three seconds following stimulus onset, statistically adjusted for prestimulus heart-rate level. Independent behavioral tests using the same stimulus conditions indicated that subjects had a mean threshold of −5 dB. In contrast to the monotonic intensity function for skin conductance, there is clear evidence of enhanced deceleratory heart-rate changes as the stimulus intensity approaches the behaviorally assessed threshold. Obviously, the response thresholds obtained for these two physiological responses are very discrepant, yet the heart-rate and behavioral thresholds are similar.

Other data reported by Jackson (1974) indicated that the deceleration obtained in response to near-threshold stimuli habituated far more slowly than did equally large decelerations elicited by stimuli well above threshold. This is just what would be expected from Sokolov's work. In the context of the typical test for an auditory threshold where it is necessary to repeat stimuli numerous times, the retardation of habituation would be a considerable advantage.

The evidence suggesting near-threshold enhancement of physiological responses is not limited to adult subjects. Several years ago we undertook a series of studies to determine if reliable estimates of hearing thresholds from individual infants could be obtained using heart-rate measures. Although this had previously been attempted in several other laboratories, some of the results were restricted to neonates (e.g., Turkewitz, Birch, & Cooper, 1972a,b) and some had employed complex stimuli that were difficult to evaluate (e.g., Eisenberg, Griffin, Coursin, & Hunter, 1964). Most had reported that pure tone stimuli were either totally ineffective (Turkewitz *et al.*, 1972a,b; Schulman, 1973) or had minimal effects (Eisenberg, 1976). Our approach was to test all infants while they were asleep so as to minimize the amount of habituation (Berg, Jackson, & Graham, 1975) and so that the infant would remain in a stable state long enough to obtain responses to several repetitions of each stimulus level. Because this research was of a preliminary nature, a variety of parameters, shown in Table 1, were explored across several studies. In all cases the stimulus was a 250-, 1000-, or 4000-Hz pure tone with a 25-msec rise/fall time. Samples A1 and A2 were part of the same experiment, differing only in whether or not subjects provided usable data for a minimum of four stimulus repetitions at each intensity. The most consistent aspect of the response elicited was an initial short-duration deceleration. Therefore, we scored the peak deceleration in the first three seconds following stimulus onset, just as Jackson had done for his adults. From the data shown in Figure 3, it is evident that, at all three frequencies, the deceleratory response is larger at 30 to 40 dB SPL than at the middle intensities, and generally larger than even for 55- to 60-dB stimuli.

Table 1. Subject and Stimulus Parameters in Each of Three Experiments on the Effects of Low Intensity Pure Tone Stimuli on Heart Rate Responses in Sleeping Infants[a]

Experiment or sample	*N*	Mean age (wks)	Stimulus intensity (dBA)	Mean ITI (sec)	Stimulus duration (sec)	No. repetitions per condition per infant
A1	10	14.4	40,45,50,55	30	2	4–8
A2	8	16.1	40,50,50,55	30	2	1–4
B	10	10.7	40,50,60	7	1	12–25
C	6	9.8	30,40,50	30	2	6

[a]Samples A1 and A2 come from the same experiment, but differ in the number of usable trails available for analysis.

The consistency across experiments of this near-threshold enhancement of response was greatest for the 250-Hz stimulus. This replicability is seen in Figure 4, which shows the data from each of the four samples separately. The response at 30 to 40 dB is greater than at 45 to 50 dB in all samples and greater than 55 to 60 dB in three of the four. At 250 Hz, the 30- to 40-dB stimuli are approximately 15 to 25 dB above that of the behaviorally determined adult free-field threshold in this environment.

Interestingly, the brief deceleratory responses seen in these sleeping infants appeared similar to the brief initial decelerations often seen in the heart-rate responses of sleeping adult subjects. Earlier we had reported data on heart-rate responses of sleeping adults to pure tones at a variety of intensities (Berg,

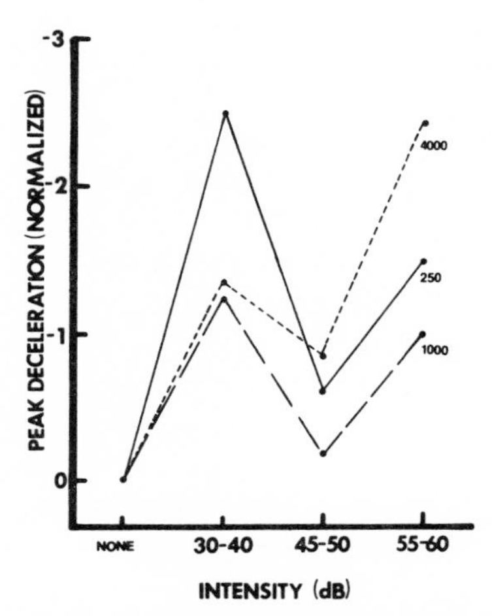

Figure 3. Points plotted are the median normalized peak deceleration of the four samples of subjects described in Table 1.

Graham, & Jackson, 1975). The response was dominated by a large acceleration, but this was consistently preceded by a small initial deceleration. In Figure 5, the amplitude of this portion of the response is plotted against stimulus intensity in the same manner as was the previous data for sleeping infants. Data are shown both for subjects tested when in sleep Stages 3 and 4 and in Stage 2 and REM sleep. Although the response changes across intensity failed to reach statistical significance, they have a function remarkably like that seen with infants: The larger deceleratory responses are produced by the low rather than middle intensity stimuli. Taken together, these several studies of both infants and adults provide compelling evidence that a component of heart-rate response is sensitive to very low intensity auditory stimuli.

More recent evidence from our laboratory suggests that the near-threshold enhancement may also be obtained using a new procedure. The procedure may be generally described as one in which the amplitude of a specific reflexive response is altered by the presentation of stimulation prior to the reflex-eliciting stimulus. In this particular case, the reflex is an eye-blink elicited with a brief puff of air directed just lateral to the outer canthus of the eye, and the modifying stimulus is a brief, pure tone presented 200 msec before the puff. Work by Reiter and Ison (1977) and others using this paradigm has demonstrated that in adult subjects, stimuli presented at behaviorally determined threshold levels are capa-

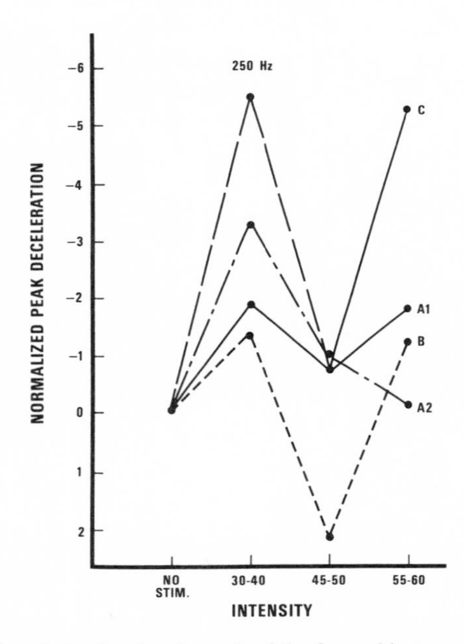

Figure 4. Normalized peak deceleration for each of the four subject samples described in Table 1.

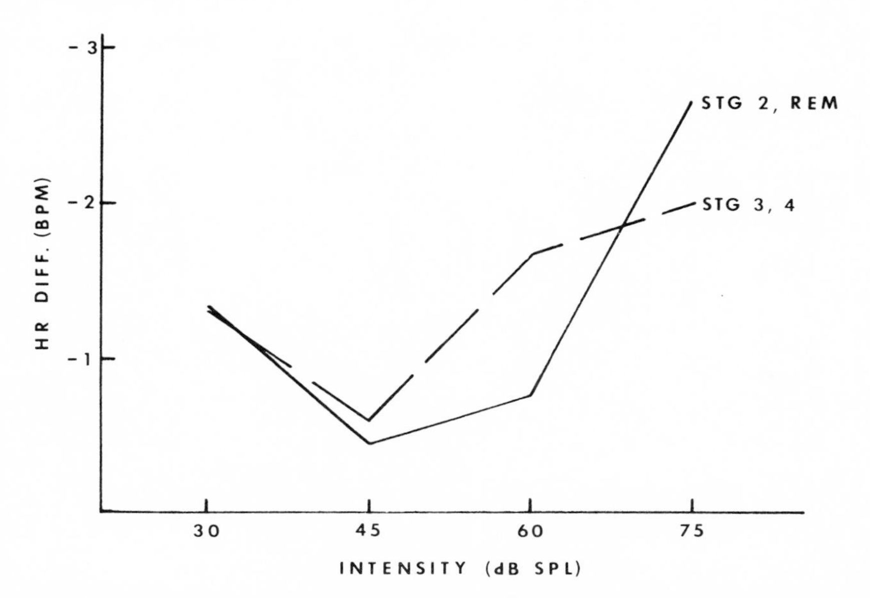

Figure 5. Adapted from Berg *et al.* (1975). STG 2, REM refers to data gathered when subjects were either in Stage 2 or Stage REM of sleep. STG 3, 4, refers to data gathered when subjects were either in Stage 3 or Stage 4 of sleep.

ble of producing a significant inhibition of the blink reflex. A wide variety of work has shown that the effects are not produced by masking, are not a conditioning phenomena, and cannot be accounted for by middle-ear reflexes (e.g., Hoffman & Ison, 1980).

The paradigm, as we employ it with infants, is shown in Figure 6. On control trials, only the 50-sec airpuff is presented and on stimulus trials one of several brief, pure tones precedes the puff by an interval of 200 msec. All babies are tested when awake. Eye blinks are recorded electrophysiologically from surface electrodes above and below one eye, and acoustic stimuli are presented via a hand-held TDH-39 earphone. Six infants of 3 to 4 months of age completed testing at 100 Hz.

The results are shown in Figure 7. The mean percent inhibition is plotted as a function of lead-stimulus intensity. That is, the mean behaviorally determined threshold of several young adults is tested with the same stimuli in the same environment. Blinks are significantly inhibited when the airpuff is preceded by the 15-dB, 1000-Hz stimuli and the 28-dB, 250-Hz stimuli. Not only do these data show the impact of near-threshold stimuli, but the 1000-Hz function suggests again that near-threshold stimuli can have an effect that is greater than more intense stimuli.

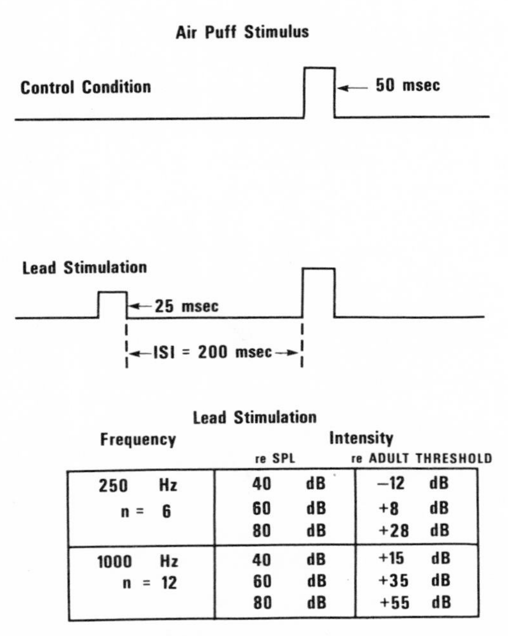

Lead Stimulation

Frequency	Intensity re SPL	Intensity re ADULT THRESHOLD
250 Hz, n = 6	40 dB	−12 dB
	60 dB	+8 dB
	80 dB	+28 dB
1000 Hz, n = 12	40 dB	+15 dB
	60 dB	+35 dB
	80 dB	+55 dB

Figure 6. A timing diagram and description of the pure tone lead stimuli used to produce blink inhibition in young infants.

How can this increased responsiveness to low-intensity stimuli be explained? Sokolov (1963) argues that such an effect occurs because near-threshold stimuli are difficult to distinguish and therefore engage more completely the orienting or attentional mechanisms. This occurs as the subject attempts to match the incoming stimuli to a developing "neuronal model" of that stimulus. In one sense, then, this notion is in agreement with that of Schneider *et al.* (1979) and Trehub *et al.* (1981); the physiological responses sensitive to low-intensity stimuli are reacting to the "significance" of such stimulation. But note that according to the Sokolovian view, the stimuli acquire this significance precisely because they have approached the limits of detectability. It would seem, therefore, that the neuronal processes governing orientation achieve an outcome for threshold-level stimuli similar to that obtained by instructions or conditioning contingencies with behavioral responses: they establish a "significance" for the stimulus that increases the likelihood of response to it. In this circumstance the "attentional threshold" and the threshold of audibility become closely convergent.

Two aspects of the data presented thus far suggest that factors other than the orienting response may be playing an important role. Contrary to what we might expect from an orienting response, the enhancement seems to occur in sleep as well as in waking states, and the affected heart-rate response is of short latency and duration, rather than the longer duration response often seen with orienting. We have argued in past reviews (Berg & Berg, 1979), as have Graham and co-workers (Graham & Slaby, 1973; Graham, Anthony, & Zeigler, 1983), that these short-duration decelerations are especially responsive to the onset characteristics of a stimulus, rather than the steady state characteristics. Similarly, Graham (1979) has argued that the reflex inhibition is primarily influenced by the transient aspects of the stimulus. If this is the case, then these responses would be ideally suited to serve as indicators of detection rather than stimulus identification. Responses that reflect stimulus detection are, of course, just what we would desire in order to measure auditory sensitivity.

There is one final implication of these data that is of practical as well as theoretical importance. A common procedure for obtaining threshold-level responding in infants is to employ a descending method of limits, that is, to initiate testing with a relatively high intensity stimulus and then to decrease intensity until responding can no longer be observed. Such a procedure presumes the presence of a monotonic function relating responsiveness to stimulus intensity. Clearly, if there is a nadir in the function at intensities above the point where near-threshold responding occurs, the descending method will seriously under-

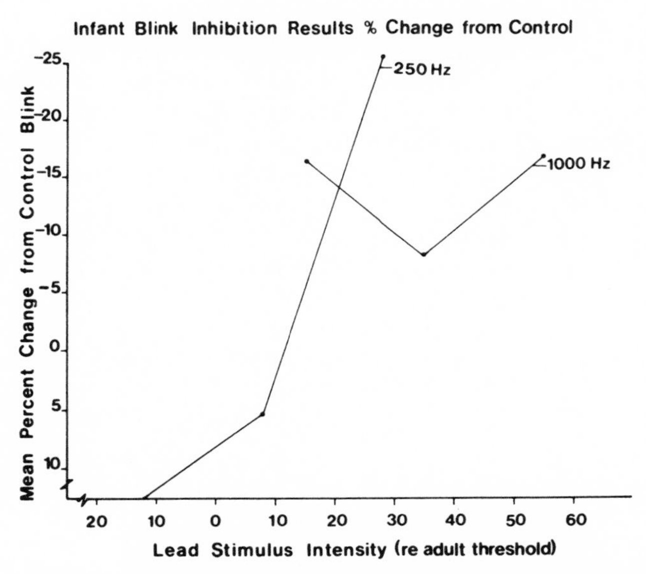

Figure 7. The mean percent change in blink response for lead-stimulus trials compared to control (no-lead-stimulus) trials. Lead-stimulus intensity is relative to behaviorally determined thresholds for the same 25-msec stimuli when presented in the same test environment as was used for the infants.

estimate the subject's auditory threshold, or at least underestimate that response system's sensitivity for detecting the threshold. Even an ascending procedure will be inaccurate unless it is started at a level below the lowest response level. In short, with physiological systems, at least, it is critical to explore a wide range of intensities and not to presume a monotonic relation between response amplitude and stimulus intensity.

As was noted at the outset of this chapter, the data on near-threshold responding seem to require a reevaluation of the view of physiological measures of auditory sensitivity. Some, but not all physiological measures, appear to be capable of sensitivities equal to or possibly even better than currently available behavioral techniques for testing auditory sensitivity in infants. The measures appear to achieve this sensitivity when they are especially sensitive to detection as opposed to the identification of stimuli. There is no *a priori* reason to expect physiological responses, as a class, to be better than or worse than behavioral responses, as a class, in reflecting the central nervous system's determination of significance. The sensitivity of any measure will probably be determined to a far greater extent by the specific nature of the paradigm in which the measure is employed, by the age and arousal state of the infant, or even by the skill and control over the subject exercised by the test personnel.

Acknowledgments

The author is indebted to Dr. Louis D. Silverstein, Dr. Marsha G. Clarkson, Margarete C. Davies, and Terry D. Blumenthal for their assistance with all phases of the research, and to Dr. Kathleen M. Berg for numerous suggestions and a careful reading of the manuscript.

References

Berg, W. K., & Berg, K. M. Psychophysiological development in infancy: State, sensory function and attention. In J. Osofsky (Ed.), *Handbook of infant development*. New York: Wiley, 1979.

Berg, W. K., Jackson, J. C., & Graham, F. K. Tone intensity and rise-decay time effects on cardiac responses during sleep. *Psychophysiology*, 1975, *12*, 254–261.

Eisenberg, F. K. *Auditory competence in early life*. Baltimore: University Park Press, 1976.

Eisenberg, R. B., Griffin, E. J., Coursin, D. B., & Hunter, M. A. Auditory behavior in the human neonate: A preliminary report. *Journal of Speech and Hearing Research*, 1964, *7*, 245–269.

Graham, F. K. Distinguishing among orienting, defence, and startle reflexes. In H. D. Kimmel, E. H. Van Olst, & J. F. Orlebeke (Eds.), *The orienting reflex in humans*. Hillsdale, N.J.: Erlbaum, 1979.

Graham, F. K., & Slaby, D. Differential heart rate changes to equally intense white noise and tone. *Psychophysiology*, 1973, *10*, 347–362.

Graham, F. K., Anthony, B. J. & Zeigler, B. L. The orienting response and developmental processes. In D. Siddle (Ed.), *Orienting and habituation: Perspectives in human research.* Sussex, England: Wiley, in press.

Graham, F. K., & Clifton, R. K. Heart rate change as a component of the orienting response. *Psychological Bulletin,* 1966, *65,* 304–320.

Hartley, H. V., & Hetrick, R. D. Ambiguities in visual identification of responses in respiration audiometry. *Journal of Auditory Research,* 1973, *13,* 305–312.

Hoffman, H. S., & Ison, J. R. Reflex modification in the domain of startle: I. Some empirical findings and their implications for how the nervous system processes sensory input. *Psychological Review,* 1980, *87,* 175–189.

Jackson, J. C. Amplitude and habituation of the orienting reflex as a function of stimulus intensity. *Psychophysiology,* 1974, *11,* 647–659.

Pool, R., Goetzinger, C. P., & Rousey, C. L. A study of the effects of auditory stimuli. *Acta Otolaryngologica,* 1966, *61,* 143–152.

Reiter, L. A., & Ison, J. R. Inhibition of the human eyeblink reflex: An evaluation of the sensitivity of the Wendt-Yerkes method for threshold detection. *Journal of Experimental Psychology: Human perception and performance,* 1977, *3,* 325–336.

Rousey, C. L., & Reitz, W. E. Respiratory changes at auditory and visual thresholds. *Psychophysiology,* 1967, *3,* 258–261.

Rousey, C., Snyder, C., & Rousey, C. Changes in respiration as a function of auditory stimuli. *Journal of Auditory Research,* 1964, *4,* 107–114.

Schneider, B. A., Trehub, S. E., & Bull, D. The development of basic auditory processes in infants. *Canadian Journal of Psychology,* 1979, *33,* 306–319.

Schulman, C. A. Heart rate auditometry. Part I. An evaluation of heart rate response to auditory stimuli in newborn hearing screening. *Neuropadiatrie,* 1973, *4,* 362–374.

Sokolov, E. N. *Perception and the conditioned reflex.* New York: Macmillan, 1963.

Trehub, S. E., Schneider, B. A., & Bull, D. Effect of reinforcement on infants' performance in an auditory detection task. *Developmental Psychology,* 1981, *17,* 872–877.

Turkewitz, G., Birch, H. G., & Cooper, K. K. Patterns of response to different auditory stimuli in the human newborn. *Developmental Medicine and Child Neurology,* 1972, *14,* 487–491.(a)

Turkewitz, G., Birch, H. G., & Cooper, K. K. Responsiveness to simple and complex auditory stimuli in the human newborn. *Developmental Psychobiology,* 1972, *5,* 7–19.(b)

COMMENTARY

Commentary on Chapters 3, 4, 5, and 6

David M. Green

Department of Psychology
University of Florida
Gainseville, Florida

Although our knowledge of infant hearing is still far from what we all desire, I think some perspective will be gained if I preface my remarks by relating a standard procedure in Scottish nineteenth-century midwifery. The Scottish midwife, once the child had been delivered and was resting comfortably, would enter the room of the sleeping infant with a large cooking pan and a wooden spoon. If the baby woke and cried when the pan was struck, the conclusion was that the infant's hearing was normal. Then, and only then, did the family start the traditional celebration for a birth in the family. Although our assessment of infant hearing is still rudimentary, we have progressed beyond the pan and the spoon.

While on the topic of assessment, let me comment on Chapter 5 by Schneider and Trehub who use behavioral means to determine the basic auditory abilities. Roughly summarized, they find in their head-orienting task that infant audiometric thresholds for pure tones are about 20 dB higher than existing adult data, except at very high frequencies. Specific tests using adults and infants show that the difference of 20 dB is observed at 10,000 Hz, but that no difference occurs at 19,000 Hz. Rather than concluding that "the major developmental changes in sensitivity will be occurring in the low-frequency region," we suggest another interpretation.

Suppose, because of attention, motivation, or misinterpretation of instruction, the infants simply do not perform as well as adults in the orienting task. That is, for reasons unrelated to their sensory capabilities, the infants score 20 dB poorer than adults in this task. According to this hypothesis, their audition sensitivity is nearly the equal of adults at all pure-tone frequencies except the very high frequencies, namely 19,000 Hz, where we must conclude that the infants are better than adults by the same 20 dB. In short, we are suggesting that the adults, due to noise exposure, or the natural course of growing older, lost about 20 dB of hearing at the highest frequencies.

This hypothesis is not as far fetched as it first might seem. There is ample evidence that hearing deteriorates with age and that it deteriorates more at higher frequencies than at lower frequencies. Of course, our most precise data come only at frequencies below about 6,000 Hz, where calibration of the headphones is possible, but there is also evidence that high-frequency hearing is even more susceptible. We need only suppose that the two adults tested by Schneider and Trehub have normal hearing at low frequencies and an audiometric loss of 20 dB at the very high, highest frequencies. We then conclude that infant hearing is "normal" at all frequencies including the highest one. Since the infants have essentially no noise exposure, we would expect them to be normal, even at these very high frequencies.

The same general, nonsensory specific differences should make the infant-masking data about 20 dB poorer than adult data, which it is.

With regard to Muir's chapter on the development of spatial sensitivity, I have nothing to contribute to the controversy concerning the functional representation of space. That discussion is better tackled by someone more perceptually oriented than I. What I do find striking and very interesting is loss of the ability to orient toward an auditory target in 2- and 3-month-old infants and the return of this ability at about the fourth month (see his Figure 2). That finding, if the data can be replicated, suggests, as Muir points out, a communality with other neonatal reflexes. Like them, one might suppose the maturation of cortical functions replacing subcortical functions. One wishes we had more data on neurophysiological development of the auditory structures, both cortical and subcortical.

Clifton's study of the precedence effect and sensory suppression of echoes also cries for more data from both the neurophysiological and the psychophysical areas. Basically, her study shows that newborns and 5-month-olds will reliably orient toward a single-sound source, whereas infants 6 to 9 weeks of age do not (replicating, in part, the results of Muir). Given two-source sounds with the one source leading the other by a few milliseconds, adults will report the sound as coming only from the leading source. The 5-month-old group orients toward the leading source, whereas the newborns show no reliable orientation, a result consistent with the hypothesis that the newborns have not developed the ability to suppress the delayed acoustic information produced by the second source. Suppression of such echoes is important in understanding auditory information in enclosed spaces where the echoes from the walls, ceiling, and floor produce reflections of the primary source and are usually unnoticed until the delays become very long. (A few hundred milliseconds delay is needed to produce a clear and distinct echo.) Tracing the time course of this change from newborns to

5-month-olds is complicated because of the unreliable behavior of the intermediate age group to a *single*-sound source. For Clifton, the single-sound source is essentially a control condition used to establish the reliability of the orienting response.

Clifton finally reports a psychophysical study for three age groups, 6-month-olds, 5-year-olds, and adults. For each group the value of the delay at which the second-sound source is first heard is measured. As might be expected, the adults first notice the second source at shorter delays than the children or infants. That adults produce smaller thresholds than infants or children does not seem particularly surprising. Better acuity for adults than children is a common finding in many areas of sensory and motor abilities. What is surprising about this result is its implications for the "inhibition" hypothesis. Since, as Clifton says, "sensory inhibition of the delayed signal produces the perception of sound localized solely on the leading side," we might reinterpret the threshold results as saying that sensory inhibition is normally quite large and extends for a long period of time in 6-month-olds. The extent of inhibition is reduced in 5-year-olds and finally reaches a low value in adults. Newborns must also have a minimal amount, since they do not orient reliably to the leading sound source in the two-speaker test. Put another way, older people will hear echoes in smaller rooms than will children. The adaptive significance of this developmental sequence is obscure.

My interpretation of this puzzle is that we need more information, both from adults and infants on the entire topic of sound localization and echo suppression—a plea endorsed by Clifton in several places in her article.

Dr. Berg's point concerning the nonmonotonicity of many physiological measures is a good one. With such indicators, and especially because habituation is a lively possibility, it is important to specify the stimulus sequence that is used to obtain the measured values. As he points out, ascending and descending methods of limits may give quite different results and one will often want to use one or the other in order to achieve some efficiency in testing. It would be interesting to see results such as those shown in Figures 1A, 3, 4, and 5 taken under conditions of ascending and descending limits. The direction of the sequence may not only modify the shape of the function but will undoubtedly change the inferred "threshold" value. Until such studies are carried out, the recommendations for audiometric procedures will be theoretical rather than practical.

PART III

CLINICAL AND DIAGNOSTIC PERSPECTIVES

CHAPTER 7

Auditory Pathologies in Infancy

George T. Mencher

Nova Scotia Hearing and Speech Clinic
Dalhousie University
Halifax, Nova Scotia,

and

Lenore S. Mencher

Co-ordinator Hearing Screening Program
Nova Scotia Hearing and Speech Clinic
Halifax, Nova Scotia

But what am I?
An infant crying in the night.
An infant crying for the light,
And with no language but a cry.

—Alfred Lord Tennyson

Auditory pathologies are generally grouped into two categories, hearing loss and central auditory dysfunction. Tennyson was certainly correct, however, in pointing out that the only language of the infant is the cry. Therefore, central auditory deficits involving auditory discrimination (ability to differentiate contrasting vowels and consonants), auditory association (ability to relate meaning to sound), auditory closure (ability to fill in missing sounds), auditory memory (ability to recall an auditory sequence), auditory localization (spatial orientation), and auditory figure–ground perception (the ability to isolate related sounds from their background) do not really appear as problems until the infant has become the toddler and language, mobility, personality, and potty training become the bane of mother's existence. Therefore, the focus of this chapter is on pathologies resulting in hearing loss. That is not to say that central auditory dysfunction does not occur, or is any less important a topic for discussion. It does say, however, that our diagnostic procedures (binaural fusion, filtered speech, alternating speech, and competing messages) require a far more sophisticated approach than can be utilized with the infant. Perhaps even more important, although it is nice to know the locus of a lesion, that knowledge will have little effect on the treatment the child and the family would receive during those early months of life, regardless of the type of auditory pathology present.

Hearing Loss

In the early 1970s, a Joint Committee on Infant Hearing was formed to evaluate the various procedures utilized to assess hearing in the newborn, and to make recommendations for appropriate procedures that could be universally applied within the United States (and Canada). The Committee consists of representatives of the American Speech-Language-Hearing Association, the American Academy of Otolaryngology—Head and Neck, the Academy of Pediatrics, and the American Nurses' Association. The Committee has issued a number of formal statements, which, coupled with the recommendations of the Nova Scotia Conference on Early Identification of Hearing Loss (Mencher, 1976), have formed the basis for most programs to identify hearing loss in infants. The primary focus of identification is through the use of ''criteria to identify infants at risk for hearing impairment,'' or what has commonly become known as the high-risk register for hearing loss (see Table 1). That register includes the major factors associated with auditory pathology in infancy and serves as the basis for this chapter. Downs and Silver (1972) recommended the use of an alphabetic mnemonic device to recall the high-risk register. The elements on the register are the same, but their sequence is altered. In this chapter, we will mirror that suggestion with some further modifications related to changes in the register.

A—Affected Family (Genetically Based Hearing Losses)

The first item on the newborn high-risk register for hearing loss, *affected family,* refers to a blood relative with a childhood hearing impairment. Clearly, this item belongs on the high-risk register because about one-half of the cases of congenital deafness are inherited (Fraser, 1971). Genetic inheritance may be through autosomal recessive, autosomal dominant, or X-linked recessive genes,

Table 1. High Risk Register for Hearing Loss

I. The criterion for identifying a newborn as at-risk for hearing impairment is the presence of one or more of the following:
   A. History of hereditary childhood hearing impairment
   B. Rubella or other nonbacterial intrauterine fetal infection (e.g., cytomegalovirus infections, herpes infection)
   C. Defects of ear, nose, or throat: malformed, low-set or absent pinnae; cleft lip or palate (including submucous cleft); and residual abnormality of the otorhinolaryngeal system
   D. Birthweight less than 1500 g
   E. Bilirubin level greater than 20 mg/100 ml serum

II. Infants falling in this category should be referred for an in depth audiological evaluation of hearing during their first two months of life and, even if hearing appears to be normal, should receive regular hearing evaluations thereafter at office or well-baby clinics. Regular evaluation is important since familial hearing impairment is not necessarily present at birth but may develop at an uncertain period of time later.

with the vast majority (approximately two-thirds) of genetic deafness arising from autosomal recessive genes.

In 1976, Bergstrom reported that of 427 patients known to be hearing impaired at a very young age, more than 40% indicated a family history of hearing impairment. Similarly, Konigsmark (1971) suggested that 35% of those congenitally deafened can trace the etiology to heredity. Bergstrom also reported that 28% of her 427 patients did not know the cause of their hearing loss. It is probable that a significant portion of those were attributable to autosomal recessive transmission as well.

Modes of Transmission

AUTOSOMAL RECESSIVE. There are at least five abnormal genes in each person that do not appear as distinct entities (Carrel, 1977). When one of these genes is located on a nonsex chromosome, that is, an autosomal recessive gene, and is matched with another autosomal recessive abnormal gene mate, the risk of producing a deaf child is one in four. Therefore, it is not surprising that in these days of low birth-rate and highly mobile families, there are many cases where normal-hearing parents produce a hearing-impaired child and can truthfully deny any family history of hearing loss. Such an occurrence is, of course, particularly critical if the hearing-impaired child is first-born, but due to the parents' lack of information about family history of hearing loss, the child would not be placed on the high-risk register and would not be screened or followed.

AUTOSOMAL DOMINANT. The genetic programming from only one parent is enough to produce dominantly inherited deafness, the risk being about 50% or one out of every two children. If both parents carry the same autosomal dominant structure, the risk is 75%. Typically, these are hearing-impaired children born to hearing-impaired parents.

X-LINKED RECESSIVE. Approximately 3% of congenital deafness is due to an X-linked recessive genetic inheritance. That is, the gene locus that determines hearing loss is on the X chromosome, one of the sex chromosomes. You may recall that females have two X chromosomes, whereas males have one X and one Y. If the male carries the deafness gene, his mate need only carry that gene on one of her two X chromosomes. In that situation, there is a 50% chance of producing a son with a hearing loss or a daughter who will be a carrier of the trait, but without a hearing loss herself. Transmission in that situation is from mother to son.

ASSOCIATED ANOMALIES. All the modes of transmission noted here can produce deafness with or without associated anomalies. In addition to a variety of outer- and middle-ear anomalies that are described in most basic texts (English, 1976; Gerber & Mencher, 1980; Jaffe, 1977; Mawson, 1967; Paparella & Shumrick, 1973), cochlear malformations may appear in three general forms. The most extreme and rare form is called the *Michel anomaly,* in which there is no inner ear and, in some cases, the entire auditory nerve may be absent. This disorder occurs in less than 1% of the profoundly deaf population. *Scheibe inner anomalies,* in which the bony labyrinth is intact but there is a loss of some of the vestibular and metabolic organs of the inner ear along with an atrophy of the organ of Corti, occur as about 70% (Bergstrom, 1976) of the congenital inner-ear anomalies. Finally, *Mondini anomalies,* which occur in 20%–30% of the cases, involve an alteration of the tissue of the cochlear duct in part or all of its contents. That is, the lower end may be abnormal, whereas the upper portion may be normal, or there may be some variation thereof.

Congenital genetic auditory disorders may also have associated anomalies in other parts of the body. These may be:

1. Integumentary or related to the skin (including albinism, piebaldness, leopardlike spots or other discoloration patterns)
2. Skeletal (including cervical or facial malformations and malformations of the ossicles)
3. Ocular (including bulging eyes, unusual coloration as seen in Waardenburg's Syndrome, and blindness)
4. Other disorders (including mental retardation, epilepsy, or psychological abnormalities)

The hearing loss that accompanies these various combinations of disorders may range from mild, to moderate, to severe, from low-frequency to high-frequency, from conductive to sensorineural, and is obviously dependent on the total syndrome involved and the extent of the expression. The reader is referred to the list of references as a starting point for further information. In any case, there is no question that a "blood relative with a childhood hearing impairment" is adequate reason for listing on the newborn high-risk register for hearing loss.

B—Breathing Difficulty (Neonatal Asphyxia)

The second major item on the neonatal high-risk register is *severe asphyxia.* In its literal definition, this is a condition whereby the brain has begun to deteriorate due to oxygen deprivation *in utero,* at or after birth. Approximately 5% of the neonatal population is affected by asphyxia, although of those, some are classified as moderate and some as severe. Moderate means that the child has required a positive pressure resuscitation of more than 1 minute. There is also a significant clinical difference between the two, in that a child suffering from severe asphyxia will probably exhibit a wide range of developmental problems, whereas one classed as moderate will essentially develop normally. Hearing loss is most often associated with severe asphyxia. In a recent statement (1981), the Joint Committee on Infant Hearing defined children as at-risk for hearing loss whose severe asphyxia was evidenced by arterial pH levels lower than 7.25, coma, seizures, or the need for continuous, assisted ventilation. The pH levels must be based on cord–artery blood at birth or no less than 20 minutes after birth, or else changes due to vigorous medical treatment may obscure the diagnosis. The neonatal brain can be deprived of oxygen through two major pathogenic mechanisms: (1) hypoxemia or diminished oxygen in the blood; and, (2) ischemia, which is a diminished blood supply to the brain tissue. Hypoxic-ischemic brain injury related to asphyxia has been called the single most important neurological problem occurring in the perinatal period and is responsible for a large percentage of the mental retardation, seizure behavior, motor deficits, and cerebral palsy seen in young children.

Hypoxemia. In hypoxemia, in addition to several important metabolic chemical changes, there is an accelerated production of lactate, a salt of lactic acid which is, in itself, the end product of the anerobic metabolism of glycogen. Initially, this production is beneficial in that it lowers the pH level, which, in turn, results in local vasodilation and increase in oxygen supply. However, as the pH levels continue to decrease and the lactate levels rise, a serious problem of tissue acidosis occurs, resulting in marked chemical changes, edema or swelling of the brain, loss of automatic regulation of the vascular system, and, of course, a parallel reduction in oxygen supply. In other words, the body mechanism chemically overreacts and compounds the problem.

Myers (1977) reported that the primate-term fetuses suffering from hypoxemia and acidosis demonstrated consistent brain swelling and cerebral infarction. Of special note was that the cerebral infarction was particularly marked in the cortex and had a "predilection for paracentral areas, especially in the posterior aspects of the hemisphere." Altenau (1975) has reported that a decrease in oxygen to the human neonate during delivery produces deficits in which the cochlear and the spiral ganglion cells appear normal. Any deafness that occurs is thought to result from damage to the cochlear nuclei. Leech and Alvord (1977) reported neuronal deficits and/or ischemic cell changes at the inferior colliculus in 85% of the perinatal-asphyxia cases they examined. Hall (1964) reported the primary loss to be in the cochlear nucleus in humans, whereas Myers (1977) and Makishima, Katz and Snow (1976) reported animal studies in which the brainstem and inferior colliculus were the primary sites of lesion. In essence, there is general agreement that the damage to the auditory system caused by hypoxemia is most likely to occur at the cochlear nucleus and/or the inferior colliculus.

Ischemia. If the patient becomes ischemic in addition to hypoxic and suffers from cerebral edema, the obstruction of blood flow to the CNS may further compound the problem. The basic difficulty is similar to hypoxia, but an additional complication called the *no reflow* phenomenon occurs (Ames, Plotkin, & Winchester, 1970). *No reflow* refers to lesions of the small vessels or capillaries, which prevent the reentry of blood into the ischemia area, even when normal pressure is restored. The swelling of perivascular astrocytes into unusual formations that project into the blood vessels and obstruct them may play a critical role in determining the reversibility of the damage to brain tissue.

Hearing Loss. Jaffe (1978) suggests that in a small percentage of cases, perinatal cerebral anoxia and asphyxia will result in high-tone, sensorineural deafness, particularly in those babies in which prematurity is an important additional factor. Robertson (1978) reported that out of 43 hearing-impaired children, 4 were severely hearing impaired, with asphyxia being the only factor, 4 were severely hearing impaired with asphyxia and low birthweight combined as factors, 4 were mildly hearing impaired with asphyxia as the only factor, and 1 was mildly impaired with asphyxia combined with low birthweight. In other words, 13 of the 43 hearing-impaired children (30%) in the study had asphyxia as a major etiological factor. In 8 of those cases, it was the only factor. Unfortunately, detailed audiograms were not presented.

With the exception of Jaffe's remarks and the Robertson paper, an extensive literature review suggests a marked absence of reports concerned with the type and pattern of hearing loss found specifically in children who suffered from perinatal asphyxia. There is material available on low birthweight, cerebral palsy, and other complications associated with perinatal asphyxia, but very little that relates exclusively to hearing loss in the asphyxiated newborn. This is particularly interesting since perinatal asphyxia is the item on the high-risk register.

A recent paper by Kileny, Connelly, and Robertson (1980) focuses on auditory brainstem responses (ABR) in perinatal asphyxia. Responses to unfiltered clicks were obtained from 14 asphyxiated neonates and a group of controls aged 3 to 17 days. The authors reported that the Wave V latency for the asphyxiated group was longer than the controls at all stimulus intensities. Further, for 60-dB clicks, both Wave I latency and Wave I to V interval were also significantly prolonged compared to the control group. Wave I latency has been associated with high-frequency, sensorineural hearing loss (Stockard & Rossiter, 1977), and an increase in that latency would appear to reaffirm Jaffe's position. Kileny, Connelly, and Robertson report that the increased Wave I to V interval was really a reflection of an increased interval at Waves I through III, something that is compatible with histopathological evidence of an edemic-based increase in intercellular space and a loss of cells in the cochlear nuclei.

One of the most significant findings of the Kileny *et al.* study, however, was that short- and long-term follow-up of several asphyxiated infants revealed both qualitative and quantitative changes in ABR that were not attributable to CNS maturation. In many situations, children progressed from total absence of response to stimuli to normal ABR patterns within 2 to 3 months. Changes were attributed to the presence of and recovery from cerebral edema or hypoxic cellular change. They stated that

> clinically evident edema as well as histologically evident hypoxic cellular change (i.e., increased intercellular space due to a disturbance in cell membrane permeability and an accumulation of fluid within the cells) are characteristic findings of hypoxic encephalopathy. When present, edema may interfere with synaptic transmission. With recovery from hypoxic encephalopathy and reduction of edema, one may expect improvement in neural transmission, provided that no permanent damage has occurred to neural tissue. (p. 157)

At the end of their follow-up study, Kileny and his co-authors reported a unilateral hearing loss in 2 of the original 14 asphyxiated neonates, the balance having normal hearing. As a result, they suggest the use of ABR to monitor progressive increases in hearing in children who have suffered from perinatal asphyxia.

It is quite evident from these results that neonatal asphyxia does interfere with the auditory process. Whether this results in permanent hearing loss, central auditory deficit, or eventually in normal audition is undoubtedly related to the degree of asphyxia, other birth complications, and, of course, genetic and environmental factors. There seems to be little doubt that neonatal asphyxia belongs on the high-risk register, but certainly further research is indicated to clarify expectations, incidence or prevalence, and extent or degree of auditory impairment.

C—Congenital Rubella, CMV and Other Infections (Viral and Other Nonbacterial Fetal Infections)

The third item on the neonatal high-risk register for hearing loss is *viral* or other *nonbacterial fetal infections;* (e.g., rubella, cytomegalovirus, herpes, toxoplasmosis, and syphilis). The two factors most often seen are, of course, rubella and cytomegalovirus, and for that reason, they will be briefly reviewed here.

Rubella. Although first described in 1752, it was not until 1941 that Gregg detailed the tetratogenic effects of the rubella virus following prenatal infection. He was also the first to link congenital cataracts and heart disease with maternal rubella. Rubella occurs in cyclical epidemics with spring as the peak season and a new epidemic occurring between 6 and 9 years after the last one. The highest incidence is usually in April and May, followed by a very sharp decline in June with the onset of school vacation for the summer (Schwartz, 1976). The highest rate of attack occurs in the 1 to 4 and the 5 to 9 age groups, with teenagers (10 to 14) being third in line. There has been a major outbreak in some part of the western world in each of the last four decades. Australia was hit in the 1940s (Fraser, 1976); Britain was the victim in the 1950s; the United States and Trinidad in the 1960s (Karmody, 1968); while Canada, Iceland, and Israel (Baldursson, Bjarnson, Halldorson, Juluisdottir, & Kjeld, 1972; Schwartz, 1976; Tibbles, Donaldson, Roy, Mencher, Goldberg, & Gibson, 1976) had epidemics in the 1970s.

MANIFESTATIONS. When an epidemic is at its height, usually 10 out of every 100 babies born show some sign of the disease (Welch, 1975). A typical rubella epidemic appeared in the Canadian Maritime Provinces (Nova Scotia, Prince Edward Island, and New Brunswick) in 1974–1975. The results of that epidemic most vividly demonstrated the devasting effects of this disease. There were 31 survivors born during an 11-month period between May, 1974, and March, 1975. Sensorineural hearing loss alone was present in only 2 cases, and cardiac defects alone were present in 2 other cases. Sensorineural hearing loss plus congenital heart disease was present in 6 cases (19.4%). In 11 cases (35%), there was a combination of heart disease plus sensorineural hearing loss, plus eye deficits. There were 5 cases (16%) with eye defects and heart disease only, while there were 3 cases (9.7%) with eye defects and sensorineural hearing loss. In other words, in 81.1% of the cases, 2 of the classic triad disorders were present. In addition to those problems, there were 9 cases of cerebral palsy and 11 (35%) of the children demonstrated a general developmental delay. The tremendous cost in diagnosis and special educational programming associated with this particular epidemic has been and continues to be phenomenal.

HISTOPATHOLOGY. Fetal infection by rubella virus generally occurs through transplacental passage and resultant lesions in the hearing-impaired child usually occur in the inner ear. Typically, these are similar to the classic membranous cochleosaccular degeneration described by Scheibe in 1892. Bergstrom (1977) reported that the pathenogenesis of the inner ear is by invasion of the stria vascularis during the viremia phase of the maternal infection. This is seen as a stria atrophy and an atrophy of the organ of Corti which results in a lack of hair cells, particularly at the basal turn of the cochlea. Brookhauser and Bordley

(1973) also reported a partial collapse of Reissner's Membrane, and adherence of that membrane to the stria vascularis and the organ of Corti following fetal rubella infection. However, in contrast to Bergstrom, they also reported that hair cells and pillar cells were plentiful in the organ of Corti and that they found no changes in the utricle, semicircular canals, or the spiral ganglia.

Richards (1964) discussed the possible relation between stapes malformation and/or fixation and maternal rubella. Similarly, the presence of only a rudimentary stapes was reported by Hemenway, Sandro, and McChesney (1969) during a histopathological study; whereas Bordley and Alford (1970) noted perivascular changes in the mucous membranes of the middle ear. In short, it is evident that rubella embryopathy is not restricted to the inner ear but may also affect the middle ear.

HEARING LOSS. The specific question in this chapter, of course, concerns hearing loss. Bordley, Brookhauser, Hardy, and Hardy, (1967) have reported that hearing loss is the single most commonly found defect in rubella children. Although there is some variation, most authors agree that the hearing loss is sensorineural, bilateral, and in some cases, enhanced by a mild conductive component (Barton & Stark, 1966; Hardy, Monif, & Sever, 1975; Miller, Rabinowitz, & Cohen, 1971). Stapes malformation or ongoing middle-ear disease, or both, usually result in an additional 10- to 20-dB hearing loss (Fitzgerald, Sitton, & McConnell, 1970; Miller, Rabinowitz, & Cohen, 1971). The audiograms are typically saucer-shaped, with the greatest loss in the middle frequencies between 500 and 2000 Hz. This is usually accompanied by poor speech discrimination. Obviously, the severity of the loss, the frequencies involved, and the extent of reduction in speech discrimination will vary from patient to patient, depending on the severity of the involvement.

In our epidemic in the Maritimes, 19 of the 31 survivors (61%) have sensorineural hearing losses. We have 12 cases of severe bilateral hearing loss (38.7%), 5 with moderate (16.1%) hearing loss, and 1 with a mild hearing loss (3.2%). There are 3 cases (9.7%) with apparent fluctuating hearing loss which, even at this stage, we have been unable to fully diagnose. There is only 1 case of a unilateral hearing loss. Sixteen of our children are wearing hearing aids (51.6%). A most interesting fact is that 29 children (93.5%) of the 31 cases have reported ongoing middle-ear pathology. In some situations, however, we believe what was originally diagnosed as middle-ear pathology is a direct result of stapes malformation or some other form of middle-ear anomaly, which impedance and delayed latency under ABR led us to believe was middle-ear disease. In several cases, the physicians have reported the ears dry at surgery. Therefore, the term *middle-ear pathology* should be interpreted to mean middle-ear difficulty, and not necessarily otitis media or some variation of that disease.

It should be noted that evidence indicates that the age of the fetus at the onset of the maternal infection appears to influence the degree of hearing loss as well. Contrary to a long-standing general belief, rubella after the first trimester can also cause hearing loss (Baldursson *et al.*, 1972). Most researchers agree that deafness is most likely to occur if the infection is in the first 16 weeks of pregnancy, but there is also a relation between maternal rubella later in pregnancy and the presence of mild to moderate degrees of hearing loss (Karmody, 1968; Ueda, Nishida, Oshima, & Shepard, 1979).

Perhaps the most critical recent information suggests that some rubella children may demonstrate progressively increasing hearing losses (Alford, 1968; Bordley & Alford, 1970; Downs, 1975; Mencher *et al.*, 1976). The deterioration begins at 6 to 9 years of age and is quite rapid. Progressivity is particularly strange when one considers the rubella disease process and its devastating effects on developing structures, which should, by all logic, stop when the disease process ends at birth or shortly thereafter. Apparently such is not the case with some unfortunate rubella children. Why some are affected and some are not is unknown. We wonder if genetic predisposition and the rubella virus combine to play a matching role in this tragic situation.

Cytomegalovirus (CMV). Cytomegalovirus is another of the fetal infections known to result in hearing loss. The disease is extremely common and easily transmitted. It is present in 1 to 1.5% of all newborns, and 80% of the population will have it by age 30 years (Marx, 1977). Transmission is usually during pregnancy via the transplacental route, or it may occur during delivery. The most classic signs include enlarged liver and spleen, jaundice, skin abnormalities and disorders similar to those seen in rubella, including hearing loss. Since hyperbilirubinemia is the most common manifestation of this disease, inclusion of CMV on the high-risk register offers something of a redundancy. Dahle, McCollister, and Stango (1974), Dahle, McCollister, and Hamner, (1979), and Gerber, Mendel, and Goller (1979) have reported severe, profound, high-frequency, and progressive sensorineural hearing loss to be associated with the disease, along with a history of chronic middle-ear problems. The similarity between CMV and rubella is rather remarkable, even histopathologically. For example, Myers and Stool (1968) reported that CMV invades the saccule, utricle, and semicircular canals, and affects Reissner's Membrane and the stria vascularis.

Herpes, Toxoplasmosis, and Syphilis. The other major nonbacterial perinatal infections (herpes, toxoplasmosis, and syphilis) also undoubtedly belong on the high-risk register. However, there seems little question that rubella is the most serious and, therefore, should head the list. Certainly, there should be further study of these latter three, and they should remain on the high-risk register. Furthermore, the number of cases seen appears to be so low as to make their inclusion on the register economically viable, even if only one child is identified.

D—Defects of Ear, Nose and Throat (Anatomical Malformation Involving the Head and Neck)

Defects of ear, nose, or throat, or anatomical malformations involving the head and neck (e.g., craniofacial syndromal abnormalities, overt or submucous cleft, morphologic abnormalities of the pinnae) represent the fourth item on the high-risk register. Many of these have already been discussed within the framework of genetically based hearing losses and need not be reiterated here. Stool and Houlihan (1977) provided perhaps the most succinct statement regarding this category:

> We assume that any child who has a marked facial deformity suffers from hearing loss until we determine that he has adequate hearing. We emphasize this because the hearing defect may be cryptic. The facial deformity is obvious, and will be of immediate concern to the parents and physicians. Auditory acuity is difficult to establish and an examination of the tympanic membranes may be difficult because the external canals are malpositioned. The loss may remain undetected until the child suffers sensory deprivation that affects subsequent language development. In infants with cleft palate, otitis media is universal. Many other severe facial deformities have been identified in patients suffering from hearing loss, especially those that involve the first and second branchial arches. (p. 43)

The incidence of hearing loss, either sensorineural or conductive, is extremely difficult to determine where anatomical malformations are concerned. For example, there is an indication of 100% occurrence of middle-ear disease in cleft palate in some reports, and as little as 15% to 20% in others. In some reports, anatomical malformation of the middle ear is blamed, whereas in others, Eustachian tube malfunction is considered the primary cause.

One of the major problems which is typical of the confusion is the question of morphological abnormalities of the pinnae. Certainly, not all cases with low-set or malformed pinnae have hearing loss. On the other hand, there are so many syndromes, in which low-set pinnae are a major symptom, that one cannot ignore that indicator as a sign of high risk for hearing loss. However, obtaining an accurate estimate of the number of children with low-set pinnae who have hearing loss is impossible, and, frankly, a meaningless statistic in terms of the real world. It should be sufficient to say that anatomical malformations of the head and neck belong on a high-risk register for hearing loss, and that further discussion on the matter is really unwarranted. But it should also be noted that the frustration associated with justifying that placement in terms of the scientific data is almost overwhelming.

E—Elevated Bilirubin (Bilirubin Level Greater Than 20 mg/100 ml Serum)

Another high-risk factor on the register for hearing loss is an *elevated bilirubin level,* sometimes referred to by the diagnostic term *hyperbilirubinemia.*

In its 1981 statement, the Joint Committee defined this as "unconjugated bilirubin exceeding 25 mg per 100 ml. serum, or exceeding the infant's weight in decagrams." The most frequent causes of hyperbilirubinemia seem to be from Rh and ABO incompatibility (erythroblastosis fetalis), although other syndromes and metabolic disorders are often associated with the problem.

Histopathology. Elevated levels of unconjugated bilirubin result from the body's inability to process and excrete bilirubin through the normal channels of urine or bile. The conjugation of bilirubin and other biological substances is provided by glucuronic acid. If there are excessive levels of these other substances or if there is an enzyme deficiency, bilirubin may not be processed adequately, resulting in toxic levels of unconjugated bilirubin. The most obvious sign of elevation of bilirubin is jaundice. Through phototherapy treatments or in more extreme cases, blood transfusions or exchanges, bilirubin can be kept below toxic levels and jaundice will disappear. Should treatment not be effective, kernicterus can occur.

Kernicterus, which literally means "nuclear jaundice," involves central lesions and degeneration of the ganglion cells. Characterized by deposits of a yellowish-orange pigment in the nuclear masses, and occasionally edema, kernicterus results in damage to the brainstem including cochlear involvement (Blakeley, 1959; Chisin, Perlman, & Sohmer, 1979; Flottorp, Morley, & Skatvedt, 1957; Jaffe, 1977; Katz, 1978). Evidence points to pathology in the cochlear nucleus and the cerebellar and basal nuclei (Blakeley, 1959; Dorland's, 1974; Goodhill, 1956). Goodhill (1956) suggests that icterus (jaundice) need not be present for lesions of the cochlear nucleus to occur. He maintains that cerebral hypoxia and anoxia can produce the same kind of pathology. Suga, Kikuchi, Hisanaga, and Takashima (1974) reported interesting findings from an autopsy performed on a neonate who died 72 hours after birth as a result of kernicterus. Marked degeneration was found in the dorsal and ventral cochlear nuclei, vestibular nuclei, the hypoglossal nuclei, and the cerebellum.

Kernicterus has also been related to the incidence of cerebral palsy (Denhoff & Robinault, 1960; Goodhill, 1956; McDonald & Chance, 1964), one form of which, *athetosis,* has been shown to correlate quite highly with hearing loss. Both are affected by blood incompatibility and hyperbilirubinemia with subsequent kernicteric brainstem lesion (Brans, 1972).

Hearing Loss. There are few reports in the literature of the exact numerical incidence of hearing loss as a result of hyperbilirubinemia. Paparella and Capps (1973) suggested approximately 7% of congenital deafness can be attributed to elevated bilirubin levels. Gerber (1977) reported a study by Stewart in which 10 of 82 infants studied had bilirubin levels greater than 20 mg per 100 ml and/or had blood-exchange transfusions. Two children had confirmed hearing loss, whereas one other was highly suspect.

Suga *et al.* (1974) described the auditory and vestibular functions of 9 children with a history of hyperbilirubinemia. Two had athetoid cerebral palsy and some hearing impairment, two had bilateral sensorineural hearing loss. Their audiograms showed a mild loss at less than 1000 Hz, sloping to severe loss at frequencies greater than 2000 Hz. The presence of a high-frequency hearing loss and athetoid cerebral palsy appears to be a consistent finding in all studies (Blakeley, 1959; Flottorp *et al.*, 1957; Kimura, 1973; McDonald & Chance, 1964). In a rather comprehensive study of the communication problems of kernicteric athetoids, Flower, Viehweg, and Ruzicka (1966) reported that in tests measuring hearing sensitivity, changes in sound levels, speech discrimination, and auditory recall, there were very few significant differences between kernicteric-athetoid, hearing-impaired children and congenitally hearing-impaired children with no Rh incompatibility. Greater differences existed within groups than across groups.

There is no doubt that the incidence of kernicteric hearing loss has decreased considerably with the development of the anti-Rh agglutenogen and phototherapy treatments. However, one of the unresolved questions is the level at which bilirubin toxicity will result in kernicterus. For example, the Joint Committee on Infant Hearing first recommended (1972) that a bilirubin level of 20 mg per 100 ml of serum, or greater, be the guideline. They recently (1981) changed that to 25 mg per 100 ml of serum. At the same time, rather than a numerical criterion, most physicians will use a formula based on gestational age by which they can compute a safe bilirubin level for each individual child. In addition, the presence of hypoxia, hypertonia, and sepsis in combination with jaundice increases vulnerability to kernicterus at even lower bilirubin levels. In any case, good medical judgment must be exercised.

The presence of jaundice is as much a warning sign as it is a diagnostic manifestation of kernicterus. Those who do not respond to treatment run a high risk of manifesting hearing loss and should, therefore, be identified as early as possible. Thus, the hyperbilirubinemia item is appropriately placed on the high-risk register. Those using the register, however, should expect very small numbers to fall within this category. In fact, within a three-year period of our study, we have only identified four children classed as at-risk because of elevated bilirubin levels. One of these failed the screening test but passed the subsequent behavioral hearing evaluation. The other infants passed the screening test. Perhaps this small incidence means we have nearly reached the point when one of the major causes of hearing loss has been virtually eliminated. We certainly hope so.

S—Small at Birth (Birthweight Less Than 1500 Grams)

Small at birth, or birthweight less than 1500 g, has been on the high-risk register since its inception in 1972. There has been some debate over whether

2000 g is a more realistic figure than 1500 g and an even more important discussion relating to the fact that the issue is not birthweight *per se,* but gestational age and birthweight combined. That is, a 40-week gestational age, 1500-g baby, is in more distress than a 33-week, 1500-g child.

There is an even more significant issue that needs to surface. That is, that all other items on the high-risk register represent causes of hearing loss; for example, genetically based hearing loss through affected family, anatomical malformation, or destruction of nerve fibers through elevated bilirubin levels, viral or bacterially based disease, or severe asphyxia. Small at birth, however, represents merely a state of the infant. A birthweight of less than 1500 g will not cause hearing loss. This is true whether the child was full-term or premature. Why, then, is this item on the high-risk register? Undoubtedly because it represents a factor that is common to other items on the high-risk register and, therefore, offers an important redundancy. This is to say, many children suffering from rubella or craniofacial anomalies or hyperbilirubinemia are low birthweight children as well. In other words, if we accept the item for its redundancy, we are justified in keeping it on the register, but not as a cause of congenital, sensorineural hearing loss. We would hasten to add here, however, that in our center, premature, low birthweight children fail the hearing screening more often than any other group, and furthermore, they show the highest incidence of middle-ear disease of any other group, except cleft palate. Furthermore, Downs (1981) has also reported that ICU babies are particularly prone to serous otitis media, with low birthweight infants being the most frequently affected group.

Within that same report, Downs has also indicated the incidence of serous otitis media in the ICU was 85%. Recently, Galambos (1978), Simmons (1977), and Stein (1981) have reported that the incidence of general hearing loss is 1 in 50 for children located in the ICU. In view of these reports and recommendations, it seems more appropriate to us that low birthweight as a redundant risk factor is not as meaningful as placement in the ICU. Clearly, the majority of high-risk infants for hearing loss will be in the ICU, so will low birthweight children not on the register, and other groups of children subjected to ototoxic drugs or head trauma, which might result in some hearing loss.

According to a previous report by Jacobson and Mencher (1981), neonatal ICU patients usually have a minimum stay of 3 days with an average of 14 days for most children. With those statistics in mind, then, we recommend the substitution of *S—Stay in the Neonatal ICU of 5 or more days,* for *S-Small at Birth* (less than 1500 g). We think the redundancy will still be there and the yield and information far more meaningful. We will be implementing that change in our center. Only time and data will tell if our judgment is correct.

F—Fetal Bacterial Meningitis

At its recent meeting in New York (1981), the U.S. Joint Committee on Infant Hearing added Neonatal Bacterial Meningitis to the high-risk register. It was argued that this particular disease process should have been added to the register a long time ago; and although it is usually understood that any child suffering from this disease would automatically be considered at-risk for hearing loss (not only because of the disease but also because of the potentially ototoxic effects of drugs used to combat it), that a firm statement was really critical to ensure follow-up of those unfortunate infants.

Manifestations. Meningitis is, of course, an inflammation of the meninges of the brain and spinal cord. Although it is often of viral origin, the most critical influence in the newborn occurs when it is bacterial in nature. The neonate with meningitis usually exhibits lethargy, anorexia, vomiting, irritability, respiratory distress and inconsistent temperature. It is believed that incompetent immune mechanisms, prematurity, and obstetrical complication (e.g., prolonged rupture of the membrane or fetal distress *in utero*) result in a susceptibility to the disease. In many cases, rapid onset allows meningitis to reach an acute stage before treatment is begun. Deafness (Overall, 1970; Teng *et al.*, 1962) may occur in the first two days of the illness, before the disease has been fully diagnosed. Among the other sequelae associated with meningitis are mental retardation, motor deficiency, and behavioral deficiency.

Meningitis is considered one of the major causes of acquired sensorineural hearing loss. It has been estimated that from 5% (Zonderman, 1959) to 27% (Bordley, Brookhauser, & Worthington, 1972) of the hearing losses in children will be due to this disease. How often this involves neonates, specifically, is not clear. Strome (1977), and Raivio (1978) report that the causes of hearing loss from meningitis may be due to:

1. Involvement of the central auditory pathway and the brainstem, thalamus, or cortex
2. Neuritis of the VIIIth nerve, resulting from exudates surrounding the nerve
3. Purulent labyrinthitis resulting from an extension of fluids into the inner ear via pathways along the VIIIth nerve or via the cochlear aqueduct
4. Septic emboli involving the vessels of the inner ear
5. Serous labyrinthitis or a nonbacterial inflammation of the labyrinth

Hearing Loss. In a study of 337 patients with a confirmed hearing loss following bacterial meningitis, Teng *et al.* (1962) found that 95% were bilateral. Of that number, 80% were considered completely deaf, while the remaining 20% had some residual hearing. Liebman, Ronis, and Lovrinic (1969) reported that the types of losses found in 49 patients with hearing impairment following meningitis included completely bilateral deafness, gradual high-tone loss, abrupt high-tone loss, and a fairly flat audiometric curve. Similarly, Rosenhall (1978) reported that 16 out of 83 patients who had suffered from meningitis developed a hearing loss. Three were bilateral to moderate, three were severe in one ear and moderate in the other, six were completely unilateral, one had a sloping, high-frequency hearing loss, and two had mild unilateral hearing losses. Gerber and Mencher (1980) reported a study of the audiograms of 28 children in special classes for the hearing impaired. The audiograms were grouped according to severity. It was found that "all the children with profound hearing losses were those who had had meningitis, and all the children who had had meningitis were in the profound hearing loss group." The only conclusion one can draw from these reports is that the incidence of hearing loss is very high and that the degree of hearing impairment may range from mild to severe, with the majority being in the severe to profound category.

Ototoxicity. No discussion of bacterial meningitis would be complete without some reference to the use of aminoglycocides or other commonly known ototoxic drugs during treatment. Seven aminoglycocides are now available for clinical use in Canada. These include neomycin, streptomycin, kanamycin, gentomycin, amikacin, tobramycin, and neotilmycin. The majority of these drugs have restricted use for specific life-threatening diseases and, of course, their ototoxicity is duly recognized. First reactions to aminoglycocides include nephrotoxicity, with neomycin the most likely to cause renal damage. The second major adverse reaction is ototoxicity, including both cochlear and vestibular involvement. These usually are reflected in tinnitus, vestibular impairment, vomiting, and hearing loss. Kanamycin, streptomycin, and amikacin are the drugs most likely to cause these problems (Hawkins, 1977).

The question of the neonate's response to ototoxic drugs is quite controversial. There appears to be no specific evidence that clearly links the use of ototoxic drugs in the newborn nursery and hearing loss in babies, although, of course, there is evidence for such a linkage in the adult. Apparently, natural immunosystems are resistant to these drugs and that resistance, coupled with good monitoring programs and limited dosage, seems to have kept this particular problem under control. For that reason, the use of ototoxic drugs *per se* is not on the newborn high-risk register for hearing loss. However, the use of ototoxic drugs, coupled with the presence of meningitis, would certainly make a child at-risk for hearing loss and should provide adequate reason for immediate referral.

Results of the Nova Scotia Newborn Hearing Screening Project

The methods and procedures utilized to screen newborns in Halifax have been presented elsewhere and need not be reviewed here (Gerber & Mencher, 1978; Mencher & Gerber, 1981). The high-risk register plays a significant role in that identification process, forming the basis for our initial screening. Since the inception of our program in 1977, we have modified that high-risk register in a number of ways. For example, we have refined the procedures to determine if a child is on the register by altering the interview questions and modifying our search through medical records. This has resulted in a decrease in the percentage of children listed under the categories of family history and hyperbilirubinemia. Another change was the recent addition of neonatal asphyxia, following the recommendations of the Saskatoon conference. As a result of these changes, the number of children seen in any one category may be subject to some slight variation at this time, compared to when our program started. However, we believe that the overall numbers presented are so great that an accurate analysis is reasonable, and any variation would really be quite insignificant.

Table 2 reflects the actual numbers of children behaviorally screened in each category. The SNCU (Special Nursery Care Unit) infants who are also on the high-risk register are classified as ''high-risk'' and were deleted from any of the other categories. The adoption, mothers's request, and miscellaneous categories constitute groups we have been studying for additional information and should generally be considered normal children.

Table 3 presents the screening failures by categories within the high-risk register and the other group. Note that there were only 79 (3.8%) failures in the ''normal group'' (adoption, mother's request, or miscellaneous), and no hearing impairment. We believe that the control group constitutes our false-positive rate. It should be noted that the false-positive rate for the normal group is very low, as is the yield (zero children). By the same token, it is obvious that the number of false positives within each of the categories of high risk is much higher, but so is the yield. We think it is significant that a greater number of high-risk children than normal children fail the hearing-screening procedure, although they may not, in fact, be hearing impaired.

Table 2. General Statistics: Nova Scotia Hearing and Speech Clinic Newborn Hearing Screening Program (Grace Hospital, 1981)[a]

Number born	16,312	
1. High risk		2222
2. SNCU		1397
3. Adoption		421
4. Mother's request		994
5. Miscellaneous		652
Total tested		5686
Number failed	293	
1.8% of those born		
5.2% of those behaviorally screened		

[a]The period covered was from January, 1978 to April, 1981; a total of 40 months.

Table 3. High Risk Register Statistics: Nova Scotia Hearing and Speech Clinic Newborn Hearing Screening Program (Grace Hospital, 1981)

Number failed	Confirmed hearing loss
A = 64	2
B = 32	4
C = 0	0
D = 17	1
E = 1	0
S = 23	9
SNCU = 81	3
Adoption = 27	0
Mother's request = 31	0
Miscellaneous = 21	0
N = 293	*N* = 19

Note that the nonhigh risk SNCU yielded 3 of the 19 hearing-impaired children (15.7%), certainly a significant enough group to consider using the SNCU on the high-risk category. However, also note that *S—Small at Birth* yielded the largest single group of hearing-impaired babies (9 or 47.3%). The combination of the two categories would certainly result in the identification of a large number of hearing losses, including the 3 in the SNCU who would have been missed if the category had not been included in our program. However, the number of false positives for the two groups combined would also be far greater than for the S group alone. That might lead to some serious criticism of the program. Thus, our own data support our arguments: Changing the high risk register by substituting *Stay in the ICU—5 days* for *S-Small at Birth,* but also clearly illustrate the disadvantages of that procedure.

In addition to the Grace Maternity behavioral-screening program, the Nova Scotia Hearing and Speech Clinic also administers a crib-o-gram and ABR hearing-screening program in the special-care nursery unit of the Izaak Walton Killam Hospital for Children. That is a regional intensive care unit and contains the most severely involved neonates in eastern Canada. Table 4 illustrates the number of children tested and those who failed either or both of the screening procedures employed. If Miscellaneous and S are combined, they form the largest single group (36.9%), thus confirming the results found at the Grace Maternity Hospital.

Table 5 represents the yield. It must be remembered that this is an ICU, and normal children are not included in the group. Therefore, it is not surprising to see a complete absence of children in group *A—Affected Family*. Groups *B—*

Table 4. General Statistics: Nova Scotia Hearing and Speech Clinic—SCNU (I.W.K. Hospital, 1981)

Number tested[a]	Number failed
A = 15	2 (7.5%)
B = 85	12 (14.1%)
C = 0	0 (0%)
D = 28	8 (28.5%)
E = 5	0 (0%)
S = 30	5 (16.1%)
Misc = 178	37 (20.8%)
N = 320	62 (19.3%)

[a]Includes multiple categories.

Neonatal Asphyxia and *D—Anatomical Defect* represent special populations because of the nature of the hospital. The majority of cleft-palate children or severe asphyxias born in Nova Scotia, outside of the metropolitan area of Halifax, are brought to this unit. Therefore, the population is skewed in that direction. The absence of rubella and hyperbilirubinemia is also significant. This undoubtedly reflects improved health care, immunization, and phototherapy. We hope that rubella and hyperbilirubinemia can, in the not-too-distant future, be deleted from the high-risk register on a permanent basis because of the total absence of cases.

Table 6 reflects the combined ICU totals for both hospitals and is offered without comment.

Table 5. Failures and Confirmed Hearing Loss: Nova Scotia Hearing and Speech Clinic—SCNU (I.W.K. Hospital, 1981)

Confirmed hearing losses[a]		
Percentage seen in UCI		Hearing impaired
7.5%	A	0
14.1%	B	6
0%	C	0
28.5%	D	9
0%	E	0
16.1%	S	3
20.8%	Misc	4

[a]Period covered was from December, 1978 to October, 1980.

Table 6. All Hospitals

Confirmed hearing losses (1978–1981)			
Category	Grace	I.W.K.	Total
A	2	0	2 (4.9%)
B	4	6	10 (24.4%)
C	0	0	0 (0.0%)
D	1	9	10 (24.4%)
E	0	0	0 (0.0%)
S	9	3	12 (29.3%)
Misc	3	4	7 (17.1%)
Totals			41

Associated "Auditory" Disorders

In 1974, we reported what we called "unusual responses" in some babies who were involved in newborn-hearing screening (Mencher & McCulloch, 1974). The four groups of responses included those with normal hearing who failed to respond to screening stimuli, those who responded only to white noise, those whose responses never habituated, and those whose responses were extreme in degree or manner. In 1976, McCulloch, Stick, and Mencher reported results from a follow-up study of some Lincoln, Nebraska, children who were classed as normal hearing, but who had failed the infant screening. These children were matched with a control group on several motor and cognitive tasks. At that time, it was hypothesized that continual failure to respond in the newborn nursery might have been due to a lag in CNS maturation and/or the presence of some central auditory disorder other than hearing loss. It was also suggested that the lag might continue and/or become more apparent as the child became older. Evidence from the study in which the control group performed better than the hearing-screening failures on every single task confirmed ongoing differences between the two groups and gave credence to our arguments.

In 1978, Mencher, Baldursson, Tell, and Levi concluded a study in Jerusalem, Israel, in which they followed a group of 106 children with normal hearing who had also failed to respond to auditory stimuli during newborn-infant screening. The children were then between 9 and 11 years of age. The children were matched with a control group. The two groups were specifically compared on a series of nine items including physical coordination, reading and mathematical skills, memory, and teachers' overall rating of school performance. Comparison was based on school and health records, test scores when available, and teacher and school administration interviews. The results were dramatic. In every single category, the normals outperformed the newborn-screening failure group. We do know that hyperactivity and lack of inhibition were the most common things noted. Whether this reflects neurological or behavioral problems or auditory disorders central to hearing loss, we cannot definitely state. Behavioral pattern was the area of largest discrepancy between the groups. Nine of the 32 children in the subject group were referred by their schools to pediatric neurology. This is prior to our study and completely independent of it. None of the 32 controls had been referred. Clearly, we need more information before we can draw any real conclusions. However, it is clear that there may be some relation between an infant's abnormal response to auditory stimulation in the newborn nursery and eventual neurological integrity.

Finally, what conclusions can we draw? First, hearing loss is the primary auditory pathology in infancy. Second, hearing screening based on the high-risk register is a reasonable approach to identifying that pathology, and third, those children who fail a hearing screening, although they may not necessarily have end-organ dysfunction, may, in fact, have central problems or central auditory problems that deserve attention.

Current procedures are not adequate to identify other auditory pathologies in infancy.

References

Alford, B. Rubella—"La bête noire de la médecine." *The Laryngoscope,* 1968, *78,* 1623–1659.

Altenau, M. M. Histopathology of sensorineural hearing loss. *Otolaryngological Clinics of North America,* 1975, *8,* 49–58.

Ames, M. D., Plotkin, S. A., & Winchester, R. A. Central auditory imperception: A significant factor in congenital rubella deafness. *Journal of the American Medical Association,* 1970, *213,* 419–421.

Baldursson, G., Bjarnson, O., Halldorson, S., Juluisdottir, E., & Kjeld, M. Maternal rubella in Iceland. *Scandinavian Audiology,* 1972, *1,* 3–10.

Barton, T., & Stark, E. Audiological findings in hearing loss secondary to maternal rubella. *Pediatrics,* 1970, *45,* 225–229.

Bergstrom, L. Viruses that deafen. In F. H. Bess (Ed.), *Childhood deafness.* New York: Grune & Stratton, 1977.

Bergstrom, L. Congenital deafness. In J. L. Northern (Ed.), *Hearing disorders.* Boston: Little, Brown, 1976.

Blakeley, R. W. Erythroblastosis and perceptive hearing loss. *Journal of Speech & Hearing Research,* 1959, *2,* 5–15.

Brans, W. A few introductory comments about hearing impairments in brain injured and especially the athetoid (Dutch text). *Tijds Logapedia Audiology,* 1972, *2,* 16–19.

Brookhauser, P., & Bordley, J. Congenital rubella deafness. *Archives of Otolaryngology,* 1973, *98,* 252–257.

Bordley, J., & Alford, B. The pathology of rubella deafness. *International Audiology,* 1970, *9,* 58–67.

Bordley, J., Brookhauser, P., Hardy, J., & Hardy, W. Observations on the effect of prenatal rubella in hearing. In F. McConnell & P. Ward (Eds.), *Deafness in childhood.* Nashville: Vanderbilt University Press, 1967.

Bordley, J. E., Brookhauser, P. E., & Worthington, E. L. Viral infections and hearing: A critical review of the literature, 1969–1970. *Laryngoscope,* 1972, *82,* 557–577.

Carrel, R. J. Epidemiology of hearing loss. In S. E. Gerber (Ed.), *Audiometry in infancy.* New York: Grune & Stratton, 1977.

Chisin, R., Perlman, M., & Sohmer, H. Cochlear and brainstem responses in hearing loss following neonatal hyperbilirubinemia. *Annals of Otorhinolaryngology,* 1979, *88,* 352–357.

Conference on newborn hearing screening, G. C. Cunningham (Ed.). Berkeley, California: Department of Health, 1971.

Dahle, A. J., McCollister, F. P., & Hamner, B. A. Subclinical congenital cytomegalovirus infection and hearing impairment. *Journal of Speech and Hearing Disorders,* 1974, *39,* 320–329.

Dahle, A. J., McCollister, F. P., & Stango, S. Progressive hearing impairment in children with congenital cytomegalovirus infection. *Journal of Speech and Hearing Disorders,* 1979, *44,* 220–229.

Denhoff, E., & Robinault, I. *Cerebral palsy and related disorders.* New York: McGraw-Hill, 1960.

Dorland's illustrated medical dictionary (20th Ed.), Philadelphia: W. B. Sanders, 1974.

Downs, M. P., Personal communication, July, 1975.

Downs, M. P., Report presented at the Joint Committee on Infant Hearing, M. Rubin, Chairman, Lexington Speech and Hearing Center, New York, March, 1981.

Downs, M. P., & Silver, H. K. The A.B.C.D.'s to H.E.A.R. Early identification in nursery, office and clinic of the infant who is deaf. *Clinical Pediatrics,* 1972, *ll,* 563–566.

English, G. R. (Ed.), *Otolaryngology.* Hagerstown, Md: Harper & Row, 1976.

Faulkner, R., & Gough, D. Rubella 1974 and its aftermath, congenital rubella syndrome. *Canadian Medical Association Journal,* 1976, *114,* 115–117.

Fitzgerald, M., Sitton, A., & McConnell, F. Audiometric, developmental and learning characteristics of a group of rubella deaf children. *Journal of Speech and Hearing Disorders,* 1970, *35,* 218–228.

Flottorp, G., Morley, D. E., & Skatvedt, M. Localization of hearing impairment in athetoids. *Acta Otolaryngologica,* 1957, *48,* 404–414.

Flower, R., Viehweg, R., & Ruzicka, W. Communicative disorders of children with kernicteric athetosis: (1) Auditory disorders. *Journal of Speech and Hearing Disorders,* 1966, *31,* 41–59.

Fraser, G. R. The genetics of congenital deafness. *Otolaryngological Clinics of North America,* 1971, *4,* 227–247.

Galambos, R. Use of the auditory brainstem response (ABR) in infant hearing testing. In S. E. Gerber & G. T. Mencher (Eds.), *Early diagnosis of hearing loss.* New York: Grune & Stratton, 1978.

Gerber, S., *Audiometry in infancy.* New York: Grune & Stratton, 1977.

Gerber, S. E., & Mencher, G. T. (Eds.) *Early diagnosis of hearing loss.* New York: Grune & Stratton, 1978.

Gerber, S. E., & Mencher, G. T., *Auditory dysfunction.* Houston: College Hill Press, 1980.

Gerber, S. E., Mendel, M. I., & Goller, M. Progressive hearing loss subsequent to congenital cytomegalovirus infection. *Human Communication,* 1979, *4,* 231–234.

Goodhill, V. Rh child: Deaf or 'aphasic'? (1) Clinical pathologic aspects of kernicteric nuclear deafness. *Journal of Speech and Hearing Disorders,* 1956, *21,* 407–410.

Gregg, N. M. Congenital cataract following german measles in the mother. *Transactions of the Ophthalmological Society of Australia,* 1941, *3,* 35–46.

Hall, J. E. The cochlea and the cochlear nuclei in neonatal asphyxia. *Acta Otolaryngologica,* 1964, Suppl. 194.

Hardy, J., Monif, G., & Sever, J. Studies in congenital rubella in Baltimore; 1964–1965. *Bulletin of the Johns Hopkins Hospital,* 1966, *118,* 97–108.

Hawkins, J. E., Jr. Ototoxicity in infant and fetus. In F. H. Bess (Ed.), *Childhood deafness.* New York: Grune & Stratton, 1977.

Hemenway, W., Sandro, I., & McChesney, D. Temporal bone pathology following maternal rubella. *Archiv Für Klinische und Experimentelle Ohren- Nasen- und Kehlkopfheilkunde,* 1969, *193,* 287–300.

Jacobson, J. T., & Mencher, G. T. Intensive care nursery noise and its influence on newborn hearing screening. *Journal of Pediatric Otorhinolaryngology,* 1981, *3,* 45–54.

Jaffe, B. F. (Ed.) *Hearing loss in children.* Baltimore: University Park Press, 1977.

Joint Committee on Infant Hearing Unpublished Statement, M. Rubin, Chairman, Lexington Hearing and Speech Center, New York, March, 1981.

Joint Committee On Infant Hearing Supplementary Statement (July, 1972). In S. E. Gerber & G. T. Mencher (Eds.), *Early diagnosis of hearing loss.* New York: Grune & Stratton, 1978.

Karmody, C. Subclinical maternal Rubella and congenital deafness. *New England Journal of Medicine,* 1968, *278,* 809–815.

Katz, J., (Ed.) *Handbook of clinical audiology.* Baltimore: Williams & Wilkins, 1978.

Kileny, P., Connelly, C., & Robertson, C. Auditory brainstem responses in perinatal asphyxia. *Journal of Pediatric Otorhinolaryngology,* 1980, *2,* 147–159.

Kimura, S. Formation of verbal behaviour in severely handicapped cerebral palsy children. *Japanese Journal of Psychology,* 1973, *44,* 97–107.

Konigsmark, B. W. Hereditary and congenital factors affecting newborn sensorineural hearing. In G. C. Cunningham (Ed.), *Conference on newborn hearing screening.* Berkeley, California: Department of Health, 1971.

Leech, R. W., & Alvord, E. C. Anoxic ischemia encephalopathy in the human neonatal period: The significance of brainstem involvement. *Archives of Neurology,* 1977, *43,* 109–113.

Liebman, E. P., Ronis, M. L., & Lovrinic, J. H. Hearing improvement following meningitis deafness. *Archives of Otolaryngology,* 1969, *90,* 470–473.

Makishima, K., Katz, R. B., & Snow, J. B. Hearing loss of a central type secondary to anoxic anoxia. *Annals of Otology,* 1976, *85,* 826–832.

Marx, J. L. Cytomegalovirus: A major cause of birth defects. *Science,* 1977, *190,* 1184–1186.

Mawson, S. R. *Diseases of the ear.* London: Edward Arnold, 1967.

McCulloch, B. K., Stick, S. L., & Mencher, G. T. The University of Nebraska neonatal hearing project—One year later. In G. T. Mencher (Ed.), *Early identification of hearing loss.* Basel: Karger, 1976.

McDonald, E. T., & Chance, G., Jr. *Cerebral palsy.* Englewood Cliffs, N.J.: Prentice Hall, 1964.

Mencher, G. T., *Early identification of hearing loss.* Basel: Karger, 1976.

Mencher, G. T., & Gerber, S. E. *Early management of hearing loss,* New York: Grune & Stratton, 1981.

Mencher, G. T., & McCullock, B. K. The Nebraska Neonatal Project. *International Audiology,* 1974, *13,* 495–500.

Mencher, G. T., Baldursson, G., Ozere, R., & Mencher, L. *Hearing loss in rubella children: An enigma.* Paper presented at the XIII International Congress of Audiology, Florence, Italy, September, 1976.

Mencher, G. T., Baldursson, G., Tell, L., & Levi, C. *Mass behavioral screening and follow-up.* Paper presented to the National Research Council, National Academy of Science, Committee of Hearing, Bioacoustics, and Biomechanisms. Omaha, Nebraska, October 1978.

Miller, M., Rabinowitz, M., & Cohen, M. Pure-tone audiometry in prenatal Rubella. *Archives of Otolaryngology,* 1971, *94,* 25–29.

Modlin, J., & Brandlin-Bennett, A. Surveillance of the congenital Rubella syndrome: 1969–1973. *Journal of Infectious Diseases,* 1974, *130,* 316–320.

Myers, R. E. Experimental models of perinatal brain damage: Relevance to human pathology. In L. Gluck (Ed.), *Intrauterine asphyxia and the developing fetal brain.* Chicago: Year Book Medical Publishers, 1977.

Myers, E. N., & Stool, S. Cytomegalic inclusion disease of the inner ear. *Laryngoscope,* 1968, *78,* 1904–1915.

Overall, J. C., Jr. Neonatal bacterial meningitis. *Journal of Pediatrics,* 1970, *76,* 499–511.

Paparella, M. M., & Shumrick, D. A. (Eds.) *Otolaryngology.* Philadelphia: W. E. Saunders, 1973.

Paparella, M. M., & Capps, M. J. Sensorineural deafness in children—Non-genetic. In M. M. Paparella & D. A. Shumrick (Eds.), *Otolaryngology—Vol. 2.* Philadelphia: W. E. Saunders, 1973.

Raivio, M. Hearing disorders after haemophilus influenzae meningitis: Comparison of different drug Regimen. *Archives of Otolaryngology,* 1978, *104,* 340–344.

Richards, C. Middle ear changes in rubella deafness. *Archives of Otolaryngology,* 1964, *80,* 48–59.

Robertson, C. Pediatric assessment of the infant at risk for deafness. In S. E. Gerber & G. T. Mencher (Eds.), *Early diagnosis of hearing loss.* New York: Grune & Stratton, 1978.

Rosenhall, U. Auditory function after haemophilus influenza meningitis. *Acta Otolaryngologica* (Stockholm), 1978, *85,* 243–247.

Rosenhall, U., & Kankkunen, A. Hearing alterations following meningitis, 1) Hearing improvement. *Ear and Hearing,* 1980, *1,* 185–190.

Schwartz, T. An extensive rubella epidemic in Israel, 1972: Selected epidemiologic characteristics. *American Journal of Epidemiology,* 1976, *103,* 60–66.

Simmons, F. B. Automated screening test for newborns: The crib-o-gram. In B. F. Jaffe (Ed.), *Hearing loss in children.* Baltimore: University Park Press, 1977.

Simmons, F. B. Diagnosis and rehabilitation of deaf newborns: Part II. *ASHA,* 1980, *22,* 475–479.

Stein, L. Report presented at a Joint Committee on Infant Hearing, M. Rubin, Chairman, Lexington Hearing and Speech Center, New York, March 1981.

Stockard, J. J., & Rossiter, V. S. Clinical and pathologic correlates of brainstem auditory response abnormalities. *Neurology,* 1977, *27,* 316–325.

Stool, S. E., & Houlihan, R. Otolaryngologic management of cranio-facial anomalies. *The Otolaryngological Clinics of North America,* 1977, *10,* 41–44.

Strome, M. Sudden and fluctuating hearing losses. In B. F. Jaffe (Ed.), *Hearing in children.* Baltimore: University Park Press, 1977.

Suga, F., Kikuchi, M., Hisanaga, S., & Takashima, Y. Kernicterus and deafness. *Otol. Fukuoka,* 1974, *20,* 22–26.

Teng, C., *et al.* Meningitis and deafness: Report of 338 cases of deafness due to cerebrospinal meningitis. *Chinese Medical Journal,* 1962, *81,* 127–130.

Tibbles, J., Donaldson, J., Roy, D., Mencher, G. T., Goldberg, M., & Gibson, E. J. Congenital rubella in the Maritimes. Meeting of the Canadian Pediatric Society, Vancouver, British Columbia: June, 1976.

Ueda, K., Nishida, Y., Oshima, K., & Shepard, T. H. Congenital rubella syndrome: Correlation of gestational age at time of maternal rubella with type of defect. *Journal of Pediatrics,* 1979, *94,* 763–765.

Welch, J. P. *Rubella Fact Sheet.* Mimeographed report. Atlantic Research Centre for Mental Retardation, Halifax, Nova Scotia, December 1975.

Zonderman, B. The preschool nerve deaf child. *Laryngoscope,* 1959, *69,* 54–89.

CHAPTER 8

The Efficacy of Brainstem Response Audiometry in the Diagnosis of Meningitis and Other CNS Pathology

P. A. Bernard

Children's Hospital
401 Smyth Road
Ottawa, Ontario

Introduction

There are two main concerns on the part of neonatalogists and pediatricians when diagnosing a sepsis neonatorum—whether or not there is a spread to meningeal spaces and what the possibilities are that a surviving infant might develop a neurological deficit. The earlier the diagnosis, the better the neurological prognosis might be.

Newborn septic state (sepsis neonatorum) refers to bacterial infection of infants during the first month of life (Siegel & McCracken, 1981). It is usually a systemic infection and occurs in 1 to 20 cases per 1000 live births depending on several predisposing factors. The most important of these is prematurity, with a maximum risk of infection if gestation is less than 35 weeks. Prematurity implies a lack of antibodies against bacterial invasion (group B streptococci, escherichia coli), a deficient supply of neutrophils, decrease in the bactericidal activity of leukocytes, impairment in the chemotaxis of neutrophils and monocytes and a deficiency of complement (C3), especially in low birthweight infants (Baker, Chalhub, & Shakelford, 1978; Christensen & Rothstein, 1980; Stoerner, Pickering, Adcock, & Morris, 1978; McCracken & Eichenwald, 1971; Schuit & DeBiasio, 1980). Various other factors may increase the incidence of systemic bacterial disease in neonates—such as delay after rupture of membranes and urinary-tract infection in the mother, both of which predispose the infant to inhalation of infected maternal secretions resulting in bacterial pneumonia and hyaline membrane formation in the lungs. Despite prompt and adequate antibiotherapy (generally utilizing a potentially ototoxic aminoglycoside), mortality remains at approximately 50%. Moreover, in approximately 30% of cases, sepsis is also complicated by patent or occult spread to meningeal spaces and brain. Neurological handicaps in surviving infants are substantial and might be prevented if diagnosis of brain injury were determined at its onset.

Unfortunately, early diagnosis of meningeal contamination in preterm neonates is very difficult as clinical signs are not specific. For instance, seizures may be a consequence of either neural inflammation or a metabolic disorder. Laboratory findings are conflicting since white blood cell count and protein content of the central spinal fluid (CSF) may be elevated in both healthy and infected preterm neonates. Hypoglycorhachia in the CSF (50% less than simultaneous blood sugar level) may arouse suspicion. The only way to demonstrate meningitis is by a positive germ finding on lumbar puncture material; however, results are usually not available for at least two days following that procedure.

Using brainstem potentials, Salamy, Mendelson, Tooley, and Chaplin (1981) demonstrated substantial differences between healthy and high-risk infants, which they attributed to insults to the central nervous system. In this chapter, we will try to demonstrate that this easy-to-do and noninvasive technique may be helpful in detecting the earliest signs of invasion of the meninges and brain by pathogenic bacteria.

Method

Subjects

Seventy-nine "normal" neonates, that is, free from sepsis neonatorum and with normal brainstem response audiometry (BSRA) recordings, were compared to 84 neonates infected with pathogenic bacteria. We considered only survivors and discounted infants with malformations, endocrinologic deficits, and infants with possible deafness from either inherited syndromes or viral infection during pregnancy (e.g., rubella, cytomegalovirus). Each group was divided into four subgroups (25–30-weeks-old, 30–35-weeks-old, 35–40-weeks-old, and > 40-weeks-old) based on true gestational age. There were, respectively, 12, 17, 24, and 26 normal infants, and 11, 24, 31, and 18 sepsis neonatorum infants in these four age categories. They did not significantly differ ($p > 0.1$) at birth when birthweight and Apgar scores were considered. All were evaluated daily by neonatalogists and most were admitted to intensive care units for at least a few days. Following discharge from hospital, they were regularly reexamined over a one-year period.

Apparatus and Procedure

All subjects were tested with BSRA technique every five days up to the time of discharge from hospital, and then one month and one year after the last recording. If meningitis was diagnosed or if the first BSRA was abnormal, recordings were repeated every two days up to the time of discharge. Because of the high humidity level in the incubators, electrical activity of brainstem nuclei was picked up with regular stainless steel EEG needles rather than with disk electrodes. We used the same procedure for healthy neonates, thus ensuring greater comparability. The active electrode was located at the vertex site with the right earlobe as reference and the left ear grounded. Far-field brainstem activity was amplified (Nicolet 200A–501A and 1007A) with 150 Hz to 3,000 Hz (−3 dB) band pass and ± 25-microvolt full-scale sensitivity. The data were averaged on-line with an 1170 Nicolet microprocessor utilizing a 40-microsecond sampling rate and 256 points per trial over a 10-msec sampling period. Artifacts induced by muscular activity were automatically rejected. Averaged traces were then displayed on an oscilloscope (Tektronix 5110), stored on magnetic, single-density floppy discettes (Nicolet 285), and plotted on a standard X–Y recorder (Hewlett-Packard 7010B). Acoustic stimuli consisted of 1.20-msec unfiltered rarefaction clicks delivered free-field through a homemade speaker with a linear response from 1,000 Hz to 16,000 Hz. Due to the presence of nasotracheal tubes, vein perfusion on the scalp, and respiration and heart-rate monitoring devices, we were obliged to use the free-field stimulation technique with the speaker located at the top of the incubator facing the right ear. Two powerful sound intensities (70 and 90 dB SPL), measured at the right ear meatus level with a Bruel and Kjaer sonometer for impulse noise, were used according to a previous study (Bernard, Pechere, & Hebert, 1980). Sound took exactly 0.4 msec to reach the pinna, and that value was later subtracted in determining latencies of evoked responses. Clicks were presented 2,048 times at a repetition rate of 9.4 sec^{-1} for each testing period. We used the Jewett and Willinston (1971) quotation but only Waves I (VIIIth nerve), III (superior olivary complex), and V (inferior colliculus area) were considered, being clearly visible in neonates. Their latencies were determined from click onset; interpeak latencies were measured from maximum positive deflection of each wave. As the infants were sometimes moving their legs or twitching, peak amplitudes were not considered, being unstable from one test to the next. We also did not attempt to use thresholds as an experimental parameter because of variations of middle-ear pressures in intubated neonates and because of the presence of mesenchyme in the middle-ear cavity in very premature infants. Correlations were done with neurological findings, blood disturbances (ionic contents, O_2 and CO_2 partial pressures) CSF cytochemistry, radiological findings, EEG recordings, and with reports of endocrinological tests. However, the investigator did not know these results until the end of the study.

Results

In controls, similar findings to those reported by Starr, Amlie, Martin, and Sandes (1977) and Schulman-Galambos and Galambos (1979) were demonstrated. Latencies of Waves I, III, and V decreased rapidly over time up to a gestational age of 35 weeks and thereafter more slowly. In neonates with a gestational age of less than 30 weeks, it was impossible to measure precisely Wave III, which was either absent or flattened.

In the "sepsis" group, responsible pathogens were escherichia coli (n = 41), group B streptocci (n = 19), listeria monocytogenes (n = 16), staphylococcus aureus (n = 4), pseudomonas aeurginosa (n = 3), and bacteroides fragilis (n = 1). This group may be divided into two subgroups depending on whether or not a bacterial growth occurred (so-called meningitis group). In the meningitis group (n = 27), all recordings were abnormal at the onset of the sepsis. Abnormality consisted primarily of a lengthening of the I to III interval (as compared to controls of similar ages). This lengthening was measured at 0.19, ± 0.03 msec and was usually associated with positive bacterial growth in the CSF two days later. Six patients had BSRA that normalized one week later, preceding the normalization of cytochemistry of the CSF and disappearance of bacteria. The normalization continued during the year of follow-up, but focal epilepsy was demonstrated with EEG recordings in all of these patients. Twenty-one neonates did not recover and the I to III interval remained significantly higher than in controls. In addition, enlargement of the III to V interval was also present in 11 of these. In all cases, CAT scanning pictures demonstrated various degrees of communicative hydrocephalus with cortical atrophy.

In the "meningitis free" group (n = 57), abnormal BSRA recordings were discovered in 39 infants during the study. Lengthening of the I to III interval could be correlated with CAT scanning pictures of hydrocephalus (n = 3), atrophy of the lower pons (n = 1), and so-called brain cysts (n = 5) corresponding to massive enlargement of ventricles. P1 hyperpositivity with flattened other waves was related to brain cysts (n = 6) and to one case of anencephaly with complete atrophy of cerebral lobes. Finally, absence of brainstem activation by sound stimuli was later correlated to bilateral profound deafness in association with other cranial nerve palsies in 6 cases. In the remaining 17 infants, abnormal recordings consisted of very poor reproducibility of waveforms with the exception of P1. We could not correlate this pattern to any disease process in the children during the year of observation. However, all had transitory episodes of apnea during their stay in the intensive care unit.

Conversely, BSRA recordings were normal in 2 cases of focal epilepsy, 5 endocrinologic disturbances mainly represented by hypopituitarism, 3 cases of gross developmental retardation, and in 2 cases of total blindness.

Discussion

In this study, we compared two groups of neonates of similar ages. One group was free of infection (n = 79), whereas the other was infected by various bacteria (n = 84). During sepsis neonatorum (systemic invasion during the first month of life) bacterial meningitis is a very insidious process. Until a positive culture is obtained from lumbar puncture (usually within 2 days), clinical signs are not specific and cytochemical changes in the CSF are difficult to interpret. Clinical judgment is most often the crucial part of the diagnostic procedure.

Since brainstem response audiometry seems to be a good indication of neural function in the lower part of the brain, it may offer a better approach for detecting disturbances in the CNS. In the present study, this procedure was regularly performed in infected neonates. Results must be discussed in conjunction with results of classical examination procedures.

One positive aspect of this technique is the possibility of very early diagnosis of meningeal space contamination since all neonates with meningitis had abnormal brainstem recordings within two days following onset of the sepsis (n = 27). These abnormalities consisted mainly of a lengthening of the I to III interval, which may be interpreted as a decreased neural velocity or a diminished neural synchronization between the cochlea and the cochlear nuclei. This is not surprising when looking at autopsy material on infants who died; cerebellopontine cisternae are usually filled with fibrinous exudate and acute edema with inflammatory infiltrates surrounding neural structures. Ventricles are also the site of acute inflammatory process.

Another advantage is the possibility of determining the course of the meningitis, since normalization of BSRA, when it occurs, usually precedes evidence of clinical and cytochemical recovery. The delay varies from one case to the next, but is never less than 4 days.

Finally, prognostication of sequelae is valuable. The absence of normalization of BSRA with a typical pattern of prolonged I to III interval and occasional hyperpositivity of these waves was seen in conjunction with hydrocephalus, which was later demonstrated by clinical observation and radiological investigation. The anatomical background might be fibrinous adhesions that entrap intracranial formations and produce shrinkage of both cranial nerves and vascular channels. However, focal epilepsy as a result of focal arachnoiditis is not predictable using the BSRA method probably because of the impossibility of testing temporal cortex function with this technique.

Much more difficult to interpret are the results of BSRA during the course of sepsis when no positive bacterial growth occurs in the CSF. BSRA failed to detect four types of alterations.

1. Focal epilepsy in all eight cases. These observations may be related to very localized cortical damage that BSRA is unable to demonstrate or to biochemical abnormalities (phenylketonuria in one case).
2. Endocrinologic disturbances. These consisted of 4 cases of hypopituitarism which usually does not affect brainstem function. One neonate was found to have hypothyroidism of congenital origin. Complete depletion of thyroid hormone is known to alter both cochlear and brain development and thus induce deafness. However, we did not attempt to determine hearing thresholds in this study, and thyroid deficit was only moderate. Its relation to sepsis is very doubtful.
3. Gross retardation was also completely unknown until 1 year of age in 3 cases. Although we should have obtained similar findings to those of Salamy *et al.* (1981), our failure to do so may be related to our choice of a control group. Many of the neonates in this group were premature and might have had abnormal development of brainstem potentials. In fact, these 3 cases of gross retardation consisted of 1 neonate with a gestational age of less than 35 weeks and 2 neonates with a gestational age of less than 30 weeks. Thus, differences between the infected neonates and "control" neonates might not be obvious.
4. Finally, two cases of blindness could not be correlated with abnormal BSRA recordings. Two slightly premature neonates (38 weeks) were found to be blind, probably due to septic neuronitis during an encephalomyelitis process.

When correlating abnormal BSRA to definite pathology of the CNS (n = 22), it is difficult to predict which type of abnormality one can expect. It may be either communicating hydrocephalus, cystlike expansions (porencephaly), dilatation of brainstem ventricles, or cranial nerve palsies. The anatomical substratum in most cases is probably either vascular occlusion or encephalitis due to the infectious process. Failure to demonstrate infection in the CSF may be a consequence of delayed lumbar puncture (after initiation of antibiotherapy), small disseminated brainstem infarcts, or disseminated intravascular coagulation (documented in 3 cases).

The 6 cases of profound bilateral deafness were associated with other cranial nerve palsies (IVth, VIth, and/or VIIth nerves) that may give rise to a suspicion of a skull base arachnoiditis process rather than an ototoxicity from the aminoglycosides generally used against the sepsis. Acute otitis media could not be demonstrated in this study. Once again, antibiotherapy may have sterilized the middle-ear cavity. Otitis media is known to occur in 15% of infants younger than 6 weeks of age or in infants placed on nasotracheal intubation for more than 7 days (Berman, Balkany, & Simmons, 1978; Shulin, Howie, Pelton, Ploussard, & Klein, 1978).

Abnormal BSRA recordings, consisting of poor reproducibility of waveforms ($n = 17$) that are uncorrelated with specific pathological conditions, are confusing. All of these babies experienced transitory episodes of apnea, but this is very common in premature neonates with or without sepsis. They may exhibit some mental retardation, infantile autism, or minimal brain dysfunction in later life. However, we must emphasize the fact that abnormal BSRA may sometimes coexist with completely normal hearing function. No ready explanation can be given for this. In such cases, both electrocochleography and cortical audiometry are normal, and behavioral audiometry does not show evidence of deafness.

Finally, it was impossible to correlate any one given pathogenic microorganism with any one given pathological finding in this study.

Conclusion

Brainstem response audiometry is thus a valuable technique in the detection of meningitis and its neurological sequelae in most cases. It is a very sensitive index of brainstem injury. During sepsis neonatorum without meningeal contamination, this technique is useful but does not pinpoint the types of pathological changes that are occurring in the CNS. Abnormal recordings within a period of one week may give rise to the suspicion of hidden brain contamination or malfunction. However, interpretation of results must be done in light of both clinical and cytochemical findings in order to avoid false positives.

References

Baker, J. P., Chalhub, L. B., & Shakelford, P. G. Ventriculitis in group B streptococcal (GBS) meningitis. *Pediatric Research,* 1978, *12,* 547 (Abstract).

Berman, S. A., Balkany, T. J., & Simmons, M. A. Otitis media in the neonatal intensive care unit. *Pediatrics,* 1978, *62,* 198.

Bernard, P. A., Pechere, J. C., & Hebert, R. Altered objective audiometry in aminoglycoside human neonates. *Archives of Otolaryngology,* 1980, *223,* 205–210.

Christensen, R. D., & Rothstein, G. Exhaustion of marrow neutrophils in neonates with sepsis. *Journal of Pediatrics,* 1980, *96,* 316–318.

Jewett, P. L., & Willinston, J. S. Auditory far field averaged from the scalp of humans. *Brain,* 1971, *99,* 681–686.

McCracken, G. H., & Eichenwald, H. F. Leukocyte function and the development of opsonic and complement activity in the neonate. *American Journal of Diseases of Children,* 1971, *121,* 120–126.

Salamy, A., Mendelson, T., Tooley, W. H., & Chaplin, E. R. Differential development of brainstem potentials in healthy and high-risk infants. *Science,* 1981, *210,* 553–555.

Schuit, K. E., & DeBiasio, R. Kinetics of phagocyte response to group B streptococcal infections in the newborn rats. *Infection Immunity,* 1980, *28,* 319–324.

Schulman-Galambos, C., & Galambos, R. Brainstem evoked response audiometry in newborn hearing screening. *Archives Otolaryngology,* 1979, *105,* 86–90.

Shulin, P. A., Howie, V. M., Pelton, S. I., Ploussard, J. H., & Klein, J. O. Bacterial etiology of otitis media during the first six weeks of life. *Journal of Pediatrics,* 1978, *92,* 893–896.

Siegel, J. D., & McCracken, G. H. Sepsis neonatorum. *New England Journal of Medicine,* 1981, *304,* 642–647.

Starr, A., Amlie, R. N., Martin, W. H., & Sandes, S. Development of auditory function in newborn infants revealed by auditory brainstem potentials. *Pediatrics,* 1977, *60,* 831–839.

Stoerner, J. W., Pickering, L. K., Adcock, E. W., & Morris, F. H. Polymorphonuclear leukocyte function in newborn infants. *Journal of Pediatrics,* 1978, *93,* 862–864.

CHAPTER 9

Electric Response Audiometry in Young Children

Hallowell Davis

Central Institute for the Deaf
818 South Euclid Street
St. Louis, Missouri

Introduction

Electric response audiometry refers to the assessment of the auditory system by means of small evoked potentials that are generated in many parts of the auditory system, from cochlea to cerebral cortex. They are separated from the background electrical activity of brain and muscle by summing, or averaging, a large number of responses. The responses are time-locked to the stimuli, and their sum increases in proportion to their number while the nearly random background largely cancels itself out. This greatly improves the signal-to-noise ratio. The technique is now quite familiar, and excellent equipment for it is commercially available.

The responses of different parts of the auditory system vary systematically from very small (microvolt) fast potentials, with latencies of a few milliseconds (msec), generated in the cochlea and the brainstem, to much larger (millivolt) slower potentials, with latencies of 100 msec or more from the cerebral cortex. We shall confine our attention to the responses of the brainstem, at or peripheral

The preparation of this manuscript was supported by a U.S. Public Health Service, Department of Health, Education, and Welfare research grant NS03856 from the National Institute of Neurological and Communicative Disorders and Stroke to Central Institute for the Deaf.

to the inferior colliculus. These responses are sensitive and reliable indicators of the function of the peripheral portion of the auditory system. The brainstem potentials will be discussed as tools for evaluation of the peripheral auditory system in infants and how they relate to the process of maturation.

Auditory Screening in the Neonatal Nursery

The best known clinical application of the auditory brainstem response (ABR) is the screening of newborn infants for possible auditory impairment. The overall incidence of such impairment in infants is of the order of 1 in 1,000 (Simmons, 1980), but in the neonatal intensive care unit (NICU), we now believe the incidence to be at least twenty times as great or of the order of 2% of surviving graduates (Roberts, Davis, Phon, Reichert, Sturtevant, & Marshall, 1982; Schulman-Galambos & Galambos, 1979; Simmons, 1980). The feasibility of routine screening of infants by ABR as they near discharge from NICU was first clearly demonstrated in 1979 by C. Schulman-Galambos and R. Galambos.

The purpose of such screening is, of course, to initiate early management of and to provide amplification for the infants with auditory impairment. It is important to begin such management early so that infants can develop, as nearly as possible, speech and language according to the normal developmental time-table. The existence of a ''critical period'' for optimal development of speech and language is now generally accepted, partly on the basis of clinical experience and partly on the basis of animal experiments on the visual system (Wiesel & Hubel, 1965). The critical period for learning speech and language is not sharply limited but it seems to center on the age span from 6 months to about 2 years. After 2 years it begins to be too late, but unless special efforts are made to identify the hearing-impaired infants, few of them are correctly recognized and appropriately managed until too late.

Total screening of all infants in maternity hospitals has been tried, using behavioral methods, but has been given up as not being cost-effective (Mencher & Gerber, 1981). The NICU, however, contains a much smaller population, and a much larger percentage are at high-risk for hearing impairment. It is therefore very logical to screen this population with the best methods available.

There are many factors that favor ABR screening as a routine procedure in the NICU. The method is simple in principle. A single auditory stimulus, an unfiltered click with a broad acoustic spectrum in the middle and upper frequencies, is employed. The intensity levels and the criteria for pass and fail can be chosen to test selectively both the high-intensity and the low-intensity response systems of the cochlea. Excellent equipment for recording ABR is commercially available, the method is noninvasive, the test can be performed during natural sleep (without sedation), the ears are tested separately, and the time consumed (½ to ¾ hour per infant) is not excessive. Furthermore, the screening by means of brainstem responses may, in principle, also reveal significant neurological defects (see Bernard, Chapter 8). Our own recent experience confirms the proposition that ABR screening is definitely feasible (Marshall, Reicher, Kerley, & Davis, 1980). The minor administrative difficulties of incorporating the test into the NICU routine and of obtaining and training competent personnel to administer it can be overcome by cooperation, good will, and an adequate budget. The initial cost of equipment is high, but if there is sufficient demand, much simpler and cheaper equipment will certainly be produced and made commercially available.

On the other hand, ABR screening in NICU to detect hearing impairment suffers from several serious difficulties and limitations. The first of these is that a failure to pass the ABR screen in NICU is only the first step in the long process of auditory habilitation. The next step is a definitive follow-up test when the infant is sufficiently mature and when it would be appropriate to begin special management if indicated. Ideally, it should be when the infant is 3- to 6-months-old. A repetition of the ABR screening test at this age is appropriate. A behavioral test (e.g., see Schneider & Trehub, Chapter 5) is still better, if the baby is cooperative. If the baby is hyperactive, suffers from multiple handicaps, is otherwise unable to cooperate, or fails the follow-up ABR screening test, the final recourse is ABR audiometry (under sedation).

Unfortunately, it is impossible in practice to get back all the babies who fail the screen in NICU. In spite of our own best efforts, 30% of the "failures" in our most recent series have been completely lost to follow-up. The experience of other investigators in the United States has been similar, although follow-up is apparently more effective in countries with governmental provision of health-care services. The major causes of our "losses" are both socioeconomic and geographic.

If at follow-up an impairment that cannot be remedied by medicine or surgery is confirmed and its severity estimated, an organization that is competent to advise and direct the parents with respect to immediate special training, the use of a hearing aid, and future educational expectations, must be available. Not every medical center can provide this combination. The success of a screening program is best stated in terms of the absolute number of children finally enrolled before it is too late in an adequate program of aural habilitation and the percentage that this represents of the living graduates of the NICU.

The largest series of infants screened by ABR in NICU, on whom such final data are available, is that of R. Galambos and his associates at San Diego. He informs me (personal communication) that, out of 890 infants tested, 141 or 15.7% failed the screen and were designated for follow-up. Of the failures, 57%

were lost. Of the 60 infants retested by the same ABR screening procedure, 13 were passed as "normal hearing" and 3 showed "neurological abnormalities only," but 10 showed a unilateral and 34 a bilateral hearing loss. Of the latter 34, 19 were judged to "probably need hearing aids," and 16 of them are "now wearing aids." These 16 represent 1.8% of the 890 total originally tested in NICU. This is a very significant overall success rate, in spite of the low percentage of retests that were obtained.

Our own much smaller series (Roberts *et al.*, 1982) shows 3 hearing-impaired infants put in contact with the Parent–Infant Program at Central Institute for the Deaf and 1 treated for otitis media. Our total number of infants satisfactorily screened is only 128, but our percentage of sensorineural impairments (2.3%) is of the same order of magnitude as in the Galambos series.

The number and percentages of successful detection and follow-up suggest that the method should be widely employed, but we (Roberts *et al.*, 1982) have serious doubts as to its cost-effectiveness. A major drawback, in addition to the difficulties and expense of follow-up, is the large number of "false failures" at the 40-dB level. Percentages are not very significant here because of the variety of follow-up tests employed and the large number lost to follow-up, but in only 3 cases was a sensorineural hearing impairment confirmed.

The large percentage of false failures is directly related to the gradual maturation of the brainstem response and of the peripheral auditory mechanism. The large majority of infants admitted to our NICU were born prematurely, with gestational ages (GA) as low as 27 weeks. The Jv wave in the ABR response to a 70-dB click was identifiable in only about half of the infants tested at postconceptional ages (PCA = GA + postnatal age) of less than 32 weeks. The Jv latency is clearly prolonged at this age, and it only gradually approaches 7.0 to 7.3 msec at 42 weeks PCA. Figure 1 shows the trend of our data (upper curve) and a similar trend in combined data taken from larger studies by Starr, Amlie, Martin, and Sanders (1977) and Galambos (1978). The curves seem to follow the logarithmic trend described by Eggermont (Chapter 2) for maturation processes in general. The displacement of our curve upward and to the right probably reflects differences in the criteria for the latency of Jv and, more importantly, for estimation of gestational age.

We have not attempted to derive a similar curve for Jv latency at 40 dB nHL for the simple reason that nearly half of our infants did not show Jv in response to this "low-level" stimulus in their final NICU test. (Our situation was complicated by the very frequent early discharge of infants from NICU, often quite abruptly, because of great demand for space in the unit.) The importance of the low-level test will be explained below in relation to the ABR audiogram. The point here is that the low-intensity response system of the cochlea matures more slowly than the high-intensity response system, and even at 40-weeks PCA there are numerous "false failures." We have decided that, in the context of the

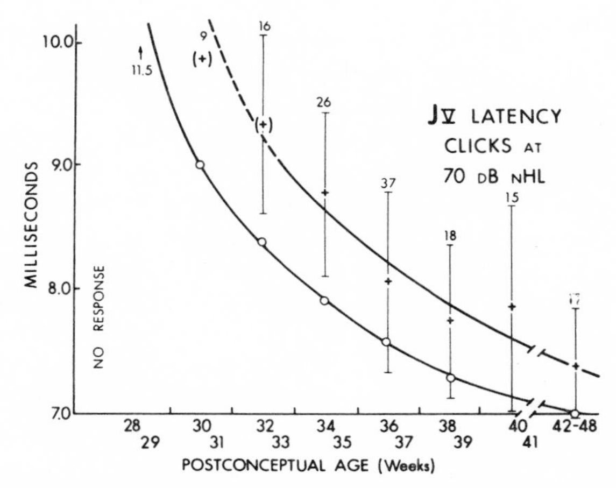

Figure 1. The curves represent the latency of the Jv wave evoked by 70-dB nHL clicks delivered at 33 clicks per second as a function of the postconceptional age (PCA) at the time of the test. The data points (circles) for the lower curve were derived from Starr *et al.* (1977) and from prepublication data kindly provided by Dr. Galambos (1978). Interpolations were necessary to arrive at a common intensity, repetition rate, and PCAs. The two sets of data were averaged. It was not feasible to estimate the standard deviations. The upper set (crosses) are from Roberts *et al.* (in press). Standard deviations (vertical lines) and the number of tests averaged for each point are indicated. Infants tested more than once in NICU contribute to 2 or 3 different sets. Tests before 34-weeks PCA, in which the latency of Jv could not be measured, are not represented, and the crosses representing the average of such incomplete sets are enclosed in parentheses. (From Roberts *et al.*, 1982, reprinted with permission.)

situation in our own hospital, continued ABR screening in NICU would not be cost-effective. Too many infants are discharged from our NICU before they are sufficiently mature for reliable ABR screening of their low-intensity auditory systems. We have chosen instead to give close attention to known high-risk factors in the NICU but to confine ABR efforts to obtaining definitive audiograms in more mature infants and young children.

The Infant's ABR Audiogram

The second established audiological use of ABR is to obtain definitive measures of auditory sensitivity in the form of a four-frequency audiogram. The method was described by Davis and Hirsh (1979), and our clinical experience with it since that time has been both extensive and satisfactory. It is the only method for accurate assessment of peripheral auditory function in children under

3 years of age and in older children who are hyperactive, emotionally disturbed, multiply handicapped, autistic, mentally retarded, unresponsive or otherwise difficult to test by behavioral methods.

The primary clinical requirement for our method is complete muscular relaxation for a period of a little over an hour. Natural sleep is very satisfactory for babies less than 3-months-old. For older children, sedation is required. We regularly employ secobarbital, but chloral hydrate is an acceptable substitute. The requirement of sedation makes the test a hospital procedure, but we do it on an out-patient basis at St. Louis Children's Hospital except for children hospitalized for other reasons.

The stimuli employed are either filtered clicks ("tone pips") or very brief tone bursts. It is necessary to make a compromise between frequency specificity (favoring long tone bursts) and synchronization of nerve impulses (favoring clicks or tone pips), as synchronization is needed to evoke a clear indicator wave in the ABR. We now are using a set of brief tone bursts at 4,000, 2,000, 1,000, and 500 Hz, each with 2-period rise and decay times and a plateau of 1 period (Figure 2). The acoustic spectra of these bursts are all similar in shape and correspond approximately to bands of noise two-thirds of an octave wide, centered at the respective frequencies (Figure 3). With such tone bursts at 4,000 and 2,000 Hz, the synchronization of nerve impulses is good enough to evoke very clear Jv responses, even at low intensities. At 1,000 and 500 Hz, however, it is necessary to use as the indicator wave a slow component of ABR that is usually removed by band-filtering of the input to the recording system. This wave, SN_{10}, is scalp-negative, with its peak at 10 msec in response to a 1000-Hz tone burst at 60 dB nHL (Figure 4). (For best recognition of this wave, we employ an input filter of the Butterworth type with low-frequency cut-off at 40 Hz and slope of 24 dB/octave. The mild distortion of waveform produced by this filter enhances the recognizability of SN_{10}.) SN_{10} does not require as close synchronization of nerve

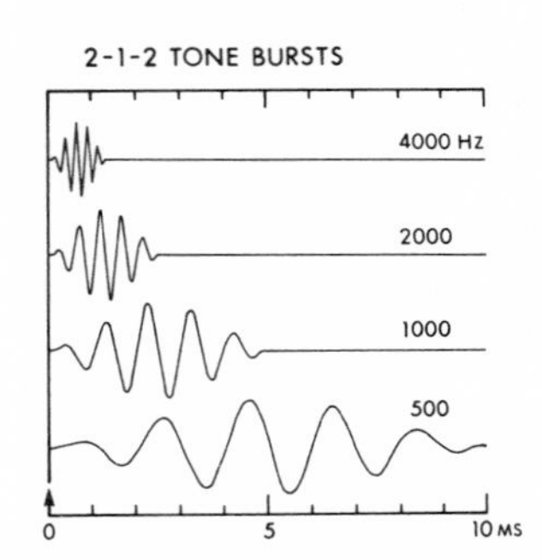

Figure 2. Tracing of waveforms of "2-1-2 tone bursts" generated by a Madsen ERA 2250 unit. Each burst has rise and fall time of 2 periods and a plateau of 1 period. Their total durations vary from 1.25 to 10.0 msec.

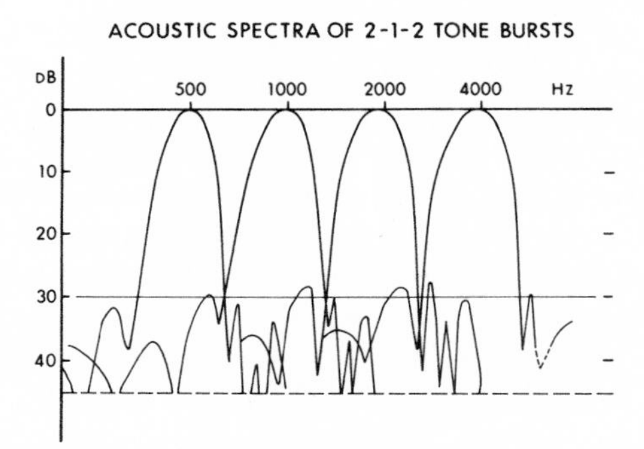

Figure 3. Acoustic spectra of the 2-1-2 tone bursts shown in Figure 2 were recorded from a standard 6-cc acoustic coupler with a Bruel & Kjaer 2203 precision sound level meter followed by a Hewlett-Packard (HP) 3581A wave analyzer set to 30-Hz bandwidth, and written out by a HP 7035B X-Y recorder. The repetition rate of the tone bursts (39/sec) produced a fine structure of equally spaced lines. The envelope was drawn by hand and was transformed (point by point) to a logarithmic frequency scale. The final curves were retraced with their peaks at the appropriate frequencies and at a common intensity reference level. Note the similarity of form of the spectra and the absence of overlap above a level 30 dB below the peaks.

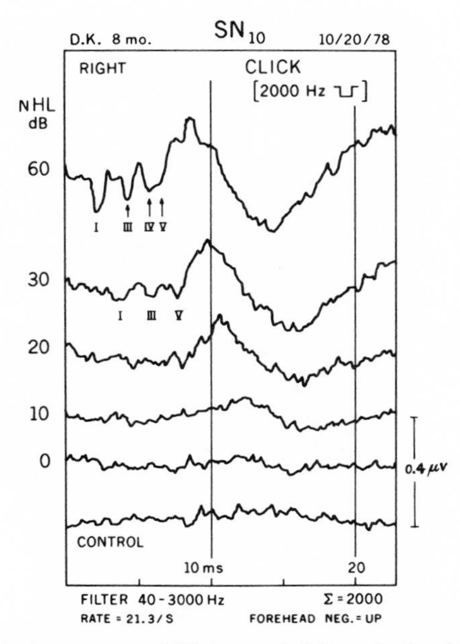

Figure 4. Slow negative brainstem wave (SN_{10}) recorded from forehead to ipsilateral earlobe in response to unfiltered clicks. Earlier, faster waves (Jewett I, III, IV, and V) are indicated. Threshold of detection of SN_{10}, by interpolation, is 5 dB nHL. (From Davis and Hirsh, 1979, reprinted with permission.)

impulses as does Jv, and, like Jv, it can be identified down to about the 10-dB sensation level in young adults.

In the audiometric test, the intensity levels are chosen according to the strategy of the game of "twenty questions," and thresholds can be estimated by interpolation (or extrapolation) on the basis of 3 or at most 4 trials. Direct validation of our audiometric estimates in young children is obviously impossible, since we test only children who cannot be tested satisfactorily otherwise. But on the basis of indirect evidence, we believe that our estimates are correct, ± 10 dB. This accuracy and the frequency selectivity shown in Figure 3 mean that we have found a very satisfactory compromise for the specification of our acoustic stimuli. Further studies of the psychoacoustic properties of these "2-1-2 tone bursts," particularly in relation to irregular audiograms, are in progress. In our opinion, however, no greater accuracy of determination is necessary for the optimum selection of a hearing aid for a hearing-impaired infant. Uncertainties concerning the shape of the audiogram need no longer delay the initiation of amplification for the hearing-impaired infant.

We have mentioned the principle of a critical period in early childhood for optimum learning of speech and language. If the analogy of animal experiments on the visual system is correct, auditory input during this period is essential for full anatomical and physiological maturation of the central auditory nervous system; and, preferably, the input should be in the form of amplified speech. Our clinical experience in the Parent–Infant Program at Central Institute for the Deaf strongly supports these inferences from animal experimentation.

Juvenile Sensory Hearing Loss

Our extensive recording of ABR in young children has revealed a type of sensory hearing loss that has hitherto gone almost unnoticed. The defect is best understood in terms of a dual mechanism in the cochlea, one part dealing with high-intensity signals and the second part with low-intensity signals.

The concept of such a dual mechanism is only gradually becoming familiar in audiology, but the two sets of responses were clearly differentiated in one of the very first papers dealing with the electrocochleogram (Yoshie, 1968). The systems differ in their thresholds, in their latencies, and in the rate of change of latency with change of intensity of the stimulus. The response of the high-intensity system has a relatively short latency, which prolongs very little with reduction in intensity. The low-intensity response is smaller and its latency at 30 dB is more than a millisecond longer than that of the high-intensity system at 60 dB, and near threshold the latency of the low-intensity response (to a click) is at least another millisecond longer still. The differences in latency are reflected in the brainstem responses, and the overall form of the ABR is different also. It is much simpler in the low-intensity range, in which Waves I, II, and III may be missing entirely and scalp-negative waves N_7 and N_9 (following Jv) may be much more prominent than the familiar P_6 (or Jv). Nevertheless, the dual mechanism itself is located in the cochlea, and its detailed nature is still unknown (Davis, 1981).

The defect in juvenile sensory hearing loss is a total impairment of the low-intensity mechanism with retention of a nearly intact high-intensity system. This type of defect was recognized clearly in the electrocochleogram of adults by Aran (1969), and was shown to be accompanied by the clinical symptom of recruitment (of loudness). The stimuli for the electrocochleograms were clicks, so the defect could only be demonstrated for the higher frequencies, that is, 2,000 Hz and above. We find in our audiograms, however, that the hearing loss extends to 1,000 and 500 Hz. The audiograms are "flat," or very nearly so, at levels ranging from 35 to 90 dB nHL. Most of the cases average between 50 and 75 dB in the better ear.

A very interesting clinical feature is that these children do not come to us because they are thought to be deaf but because of language difficulties such as slow learning or very defective speech. They respond well to loud sounds but they do not hear soft sounds. This pattern is fully consistent with recruitment, and the language defects probably reflect the fragmentary, erratic, inconsistent "speech" that they perceive.

We call the condition a sensory (not sensorineural) loss because the defect seems so clearly to lie in the cochlea. The low-intensity system, which almost certainly involves the outer hair cells, is not functioning. The high-intensity system may give a completely normal ABR at 60 dB nHL and higher, or perhaps the amplitude is somewhat reduced. The latencies are short and stable. Actually the normal threshold of the high-intensity system seems to be about 45 dB nHL, but it may be moderately elevated in this recruiting type of sensory impairment.

This sensory impairment is most clearly distinguished from a conductive impairment by the absence near threshold of any Jv wave with its long latency of 8 msec (or more). The latency of the high-intensity response is still short when the amplitude goes to zero. We shall soon publish elsewhere a more complete description and discussion of this cochlear defect. We mention it here because it explains why we were concerned to include a low-intensity stimulus in the ABR screening study in NICU. We had realized that juvenile sensory hearing loss is frequently congenital. The birth histories of some of our early cases had included very severe perinatal asphyxia and actually our two youngest cases were correctly detected as failures in our ABR screening.

Perspective

This discussion of the audiological applications of ABR in infancy has been a survey of several recent clinical studies and has included a rather unfamiliar interpretation of the cochlear mechanism. The topics are not logically connected with one another in a single sequence but all relate in one way or another to the maturation of the auditory system, either peripheral or central. Some of the more important conclusions and opinions that emerge are the following.

1. Early detection, accurate assessment, and effective management of hearing-impaired children before the end of a "critical period" are essential for the optimal exploitation of residual hearing and thereby the relatively normal development of speech and language. The critical period is a phase in the maturation of the central nervous system which probably centers on the age range from 6 months to 2 years.

2. A significant proportion of infants in NICU have a sensorineural hearing impairment.

3. It is technically feasible to detect the hearing-impaired infants by a simple screening procedure based on ABR.

4. The interpretation of ABR screening tests in NICU is confounded by the maturation of the peripheral auditory mechanisms. Reliable responses to high-intensity (70dB nHL) clicks appear at postconceptional ages (PCA) of about 30 weeks, and they approach mature form and latency at about 40 weeks. A second more sensitive and more vulnerable mechanism matures more slowly, so that responses to low-intensity (30 or 40 dB nHL) clicks often do not appear until about 38 weeks PCA.

5. The exact nature of the low-intensity mechanism is unknown, but its absence causes a sensory hearing loss of recruitment type with a "flat" audiogram, usually at 50 to 75 dB nHL.

6. A large proportion of injuries to the low-intensity system are congenital. The cases can be detected by ABR screening in NICU but only at the expense of many false failures. This is because many infants are discharged from NICU while their peripheral auditory mechanisms are still immature.

7. Follow-up tests on graduates of NICU who fail the ABR screen are essential, but it is difficult in practice to realize a follow-up rate better than 50%.

8. These difficulties, the costs of equipment, and the administrative problems mean that ABR screening in NICU is *not* cost-effective at present in many actual hospital situations (including our own), particularly in the United States.

9. With slightly modified commercial equipment it is now easy to obtain by ABR a four-frequency audiogram (500, 1,000, 2,000, and 4,000 Hz) on any child at any age, but a basic requirement of the method is sedation. Complete muscular relaxation is necessary.

10. The ABR audiometric method tests only the auditory mechanism peripheral to the inferior colliculus. It is probably accurate to ± 10 dB in predicting behavioral thresholds and it is sufficiently frequency selective for the selection of an optimal hearing aid. The method is ideal for follow-up of infants who fail the ABR screening test and also for correct diagnosis and evaluation of juvenile sensory hearing loss at any age.

References

Aran, J. M. L'électrocochléogramme: Méthodes et premiers résultats chez l'enfant. *Revue de Laryngologie Otologie Rhinologie* (Bordeaux), 1969, *90*, 615–634.

Davis, H. The second filter is real, but how does it work? *American Journal of Otolaryngology*, 1981, *2*, 153–158.

Davis, H., & Hirsh, S. K. A slow brainstem response for low-frequency audiometry. *Audiology*, 1979, *18*, 445–461.

Galambos, R. Use of the auditory brainstem response (ABR) in infant hearing testing. In S. E. Gerber & G. T. Mencher (Eds.), *Early diagnosis of hearing loss*. New York: Grune & Stratton, 1978.

Marshall, R. E., Reicher, T. J., Kerley, S. J., & Davis, H. Auditory function in newborn intensive care unit patients revealed by auditory brainstem potentials. *Journal of Pediatrics*, 1980, *96*, 731–735.

Mencher, G. T., & Gerber, S. E. (Eds.). *Early management of hearing loss*. New York: Grune & Stratton, 1981.

Roberts, J. L., Davis, H., Phon, G. L., Reichert, T. J., Sturtevant, E., & Marshall, R. E. Auditory brainstem responses in pre-term neonates: Maturation and follow-up. *Journal of Pediatrics*, 1982, *101*, 257–263.

Schulman-Galambos, C., & Galambos, R. Brainstem evoked response audiometry in newborn hearing screening. *Archives of Otolaryngology*, 1979, *105*, 86–90.

Simmons, F. G. Patterns of deafness in newborns. *The Larynogoscope*, 1980, *90*, 448–453.

Starr, A., Amlie, R., Martin, W. H., & Sanders, S., Development of auditory function in newborn infants revealed by auditory brainstem potential. *Pediatrics*, 1977, *60*, 831–839.

Wiesel, T. M., & Hubel, D. H. Extent of recovery from the effects of visual deprivation in kittens. *Journal of Neurophysiology*, 1965, *28*, 1060–1072.

Yoshie, N. Auditory nerve action potential responses to clicks in man. *The Laryngoscope*, 1968, *78*, 198–215.

COMMENTARY

To BER or Not to BER

That Is the Question

J. J. Eggermont

Department of Medical Physics and Biophysics
University of Nijmegen
Nijmegen, The Netherlands

Brainstem-electric-response (BER) testing in the neonate and young child can be divided into screening procedures and audiometric procedures. Screening comprises the detection of brainstem abnormalities as well as hearing dysfunction; audiometric testing comprises the quantification of hearing loss for a sufficient number of audiometric frequencies. Screening is needed in the very young; audiometric testing by way of BER is necessary in the difficult-to-test, that is, those children who cannot be tested reliably by behavioral means.

As Davis (Chapter 9) has shown, screening in the intensive care unit (ICU) appears to detect (later confirmed) hearing failures in 2% to 5% of the number of neonates tested. This holds for his own data as well as for the considerably larger group studied by Galambos and his associates (e.g., Schulman-Galambos & Galambos, 1979) and indicates that screening in the ICU is feasible, although not necessarily cost effective. Mencher and Mencher's (Chapter 7) extensive study of the relation between entry on the high-risk register and the probability of having a hearing disorder reveals that for children on the high-risk register who are not ICU cases, 5% of over 5,500 cases failed the screening test, but hearing loss was confirmed subsequently in only 19 cases. In the ICU population, Mencher and Mencher's studies point to 22 cases of confirmed hearing loss out of 320 (7%). Extensive behavioral testing was carried out in this group with the crib-o-gram test and the BER-test in its audiometric form, that is, testing with tone pips at 0.5, 1, 2, and 4 kHz. Bernard's (Chapter 8) study indicates the utility of BER in testing newborns in septic states for early diagnoses of possible brain injury, especially signs of invasion of the meninges and brain by pathogenetic bacteria.

All three studies face the following basic problems centered around questions concerning the number of false positives and false negatives:

1. Does an abnormal BER exist while hearing function is normal?
2. Does a normal BER exclude abnormal hearing function?
3. Does the BER help to document the types of pathological changes that are occurring?

There are also three additional questions involving treatment that must be considered. Will the early intervention permitted by early diagnosis of brainstem abnormalities in the ICU be decisive for treatment? Second, will early diagnosis of hearing abnormalities improve the success of rehabilitation? Finally, is the application of the BER cost-effective?

Before we attempt to answer these questions a very intriguing and also frightening observation of Mencher and Mencher needs to be recalled. Children who fail the screening procedure may not have end organ lesions but may have associated ''auditory'' disorders, that is, they may continue to have a lag in CNS maturation and have poorer performance than their ''normal'' classmates. Considering this observation and taking into account the ease of testing BER even in a noisy (electrically as well as acoustically) environment as is the ICU, one must be inclined to test when necessary. For this purpose, a rigid application of a selection criterion might be helpful. We should also take into account that BER screening may eliminate a number of other tests which are more invasive, such as central spinal fluid tests by lumbar puncture. This certainly should be included in calculating the cost-effectiveness of the BER-test.

Screening tests are usually done with click stimuli; audiometric testing has to be done with some type of frequency specific stimulation. Too often screening procedures are used for audiometric testing; a click-threshold and a Wave V latency–intensity function are not a substitute for an audiogram. They are related to it, but in a very intricate manner. Given the audiogram and the type of hearing loss, one can predict the click threshold and the course of the latency–intensity function. However, successful prediction in the reverse direction is highly doubtful.

Davis (Chapter 9) uses tone pips, having two periods of sine wave for rise and fall and one period as a ''plateau,'' as stimuli in his audiometric procedure. This makes the toneburst duration frequency-dependent. This type of stimulation has been advocated and used extensively in electrocochleography by Eggermont, Odenthal, Schmidt, and Spoor (1974). Its use for audiometric evaluation in young children has been reported to be accurate as well as reliable (Eggermont, 1976; Spoor & Eggermont, 1976). One expects, therefore, similar results using

BER, with the provision that the signals are much smaller than in electrocochleography, and testing duration therefore will be longer. To increase testing efficiency, Davis uses Wave V for the 2- and 4-kHz tone pips. This certainly works. Others advocate the use of the middle-latency responses for the lower tonepip frequencies and the BER for the higher frequencies (Terkildson, Osterhammel, & Huis in't Veld, 1978). This indicates that especially the threshold estimation at 500 Hz poses some problems. Another method uses click stimulation in the presence of a high-passed noise masker with various cut-off frequencies; this method of selective masking appears to be quite reliable in evaluating audiograms (Don, Eggermont, & Brackman, 1979). Problems with this latter method will arise when hearing remnants have to be detected, that is, when thresholds at 500 Hz exceed 70 dB HL. All methods using clicks as stimuli will suffer from this; the low-frequency energy in the click may not be sufficient to reach threshold when the higher frequencies in the click have been masked by the high-pass masker. Tone burst studies generally do not have this drawback, but, on the other hand, do not generate well-synchronized BER-waves. Thus, the method applied by Davis is a way out of this trap.

Audiometric BER testing requires much more effort than click-screening, but it is used generally at a critical stage in the development of the child where rehabilitation can no longer wait. Furthermore, increasing knowledge about the effects of sound deprivation on CNS maturation combined with the observations of Mencher and Mencher on the "lag" that ICU children continue to have at later times of life, may shift the age at which "rehabilitation can no longer wait" more and more into the neonatal range. How to balance the more extensive testing required against possible prevention of setbacks due to hearing impairment is a difficult but necessary component of a cost/benefit analysis. Improved knowledge about the auditory brainstem response and its underlying mechanisms will certainly aid in coming to a reasonable decision as to whether or not brainstem electric response procedures should be employed.

References

Don, M., Eggermont, J. J., & Brackman, D. E. Reconstruction of the audiogram using brainstem responses and high-pass noise masking. *Annals of Otology, Rhinology and Laryngology,* 1979, *88,* Suppl. 57, 1–20.

Eggermont, J. J. Electrocochleography. In W. D. Keidel & W. D. Neff (Eds.), *Handbook of sensory physiology* (Vol. V/III). New York: Springer, 1976.

Eggermont, J. J., Odenthal, D. W., Schmidt, P. H., & Spoor, A. Electrocochleography: Basic principles and clinical application. *Acta Oto-laryngologica. Supplement* (Stockholm), 1974, 316, 1–84.

Schulman-Galambos, C., & Galambos, R. Brainstem evoked response audiometry in newborn hearing screening. *Archives of Otolaryngology,* 1979, *105,* 86–90.

Spoor, A., & Eggermont, J. J. Electrocochleography as a method for objective audiogram determination. In S. K. Hirsh, D. H. Eldredge, I. J. Hirsh, & S. Silverman (Eds.), *Davis and hearing: Essays honoring Hallowell Davis.* St. Louis: Washington University Press, 1976.

Terkildsen, K., Osterhammel, P., & Huis in 't Veld, F. Recording procedures for brainstem potentials. In R. F. Naunton & C. Fernadez (Eds.), *Evoked electrical activity in the auditory nervous system.* New York: Academic Press, 1978.

PART IV

AUDITORY PATTERN PERCEPTION

CHAPTER 10

Auditory Pattern Perception in Infancy

Sandra E. Trehub

Centre for Research in Human Development
Erindale College, University of Toronto
Mississauga, Ontario, Canada

The study of auditory perception has been primarily the study of single sounds, and the study of pattern perception, that of visual patterns. Consider, instead, *sequences* of sounds as well as *sets* of single sounds, focusing not on acuity but on the perception of structure in these complex auditory patterns. For the present purposes, structure will refer to the relations between elements of a sound pattern or the relations between different members of a set of sounds. In this context, I will review the very limited research on auditory pattern perception in infancy and present some findings from our own laboratory.

Our own work and our consideration of the work of others has been guided by certain general principles derived mainly, but not entirely, from the study of visual perception. These principles appear to be relevant to adult audition, notwithstanding the fact that visual patterns are necessarily articulated in space, and auditory patterns, in time. For example, perceptual systems, both auditory and visual, are viewed as searching for regularities or structure in the environment, and this structuring or segmentation of the perceptual world is thought to underlie all cognitive processing (Garner, 1962, 1974; Jones, 1978; Koffka, 1935; Neisser, 1967).

There may be alternative structures or organizations that are applicable to, or perceptible in, a pattern or set of patterns, but there is evidence to suggest that some organizations are more "natural" or prominent than others. Consider the

The research reported in this chapter was assisted by grants from the Natural Sciences and Engineering Research Council of Canada and from the University of Toronto.

well-known form depicted in Figure 1, which is perceived as a triangle with superimposed bar, rather than the unnatural alternative with missing elements. In analogous fashion, an extended tone with a gap filled by noise is heard as a continuous tone with superimposed noise (Bregman, 1978c). Similarly, the repeated alternation of loud and weak signals may be heard as a continuous weak signal with intermittent loud signal superimposed (Bregman & Dannenbring, 1977; Warren, Obusek, & Ackroff, 1972). The nature of such organization will vary, depending on the characteristics of the stimulus and the context of stimulation. For example, perception of the temporal order of sounds poses few problems in rapidly presented sequences of speech (Foulke & Sticht, 1969) or music (Winckel, 1967) but may be problematic with other sounds (Bregman & Campbell, 1971; Broadbent & Ladefoged, 1959; Warren, Obusek, Farmer, & Warren, 1969). In line with several theorists (e.g., Attneave, 1954; Bregman, 1978c, 1981; Pomerantz, 1981), such structuring processes are seen as operating to maximize economy and coherence in a complex world, directing attention toward recurrent and, therefore, meaningful relations and away from fortuitous cooccurrences.

Although the infant might be expected to have some propensities to structure or organize environmental input, it is not at all obvious how extensive such

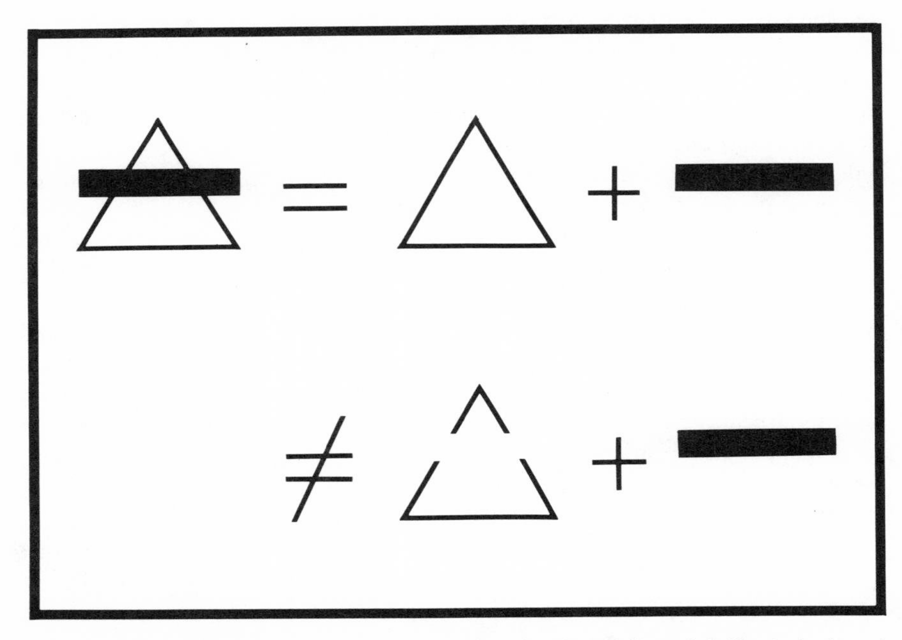

Figure 1. The figure with bar is typically perceived as a triangle partially occluded by the horizontal bar.

propensities might be or how they might be mapped onto the perception of auditory patterns, given the infant's relative inexperience with environmental regularities. It is even less obvious how one would ascertain infants' strategies for the processing of auditory patterns. Instead of the adult subject who can provide verbal descriptions, compare stimuli systematically, make qualitative and quantitative judgments, and otherwise furnish precise data in a wide variety of tasks, the infant can only provide indirect information by his or her ability to discriminate one stimulus from another. Discrimination tasks build on the infant's rather limited response repertoire, which includes a generalized orienting response, crying, habituation of the orienting response, more crying, and localization of laterally presented sounds, to name the prominent behavioral responses.

The most common infant discrimination procedures in current use are two rather primitive versions of the same–different task. One is the habituation–dishabituation design, which provides for repeated presentation of a stimulus until the infant's response (e.g., heart-rate deceleration, sucking) wanes, followed by presentations of a novel or changed stimulus (Chang & Trehub, 1977a,b; Leavitt, Brown, Morse, & Graham, 1976; Trehub, 1973, 1976). The expectation here is that the response will be reinstated or increased if the change is perceived. The other procedure involves training an infant to respond (e.g., by head turning) to some change in auditory stimulation (e.g., Eilers, Wilson, & Moore, 1977; Trehub, Bull, & Thorpe, 1984). In this case, the infant is encouraged to ignore (i.e., not respond to) a repeating background stimulus and reinforced for responding only during brief intervals when the background stimulus is replaced with a novel stimulus.

With respect to auditory pattern perception, the simplest substantive question that one can ask of infants is whether they can discriminate different patterns. Several researchers have reported that infants can distinguish contrasting sequences of 6 or 8 tones (Kinney & Kagan, 1976; McCall & Melson, 1970; Melson & McCall, 1970), but in these studies the different patterns invariably had different initial tones, raising the possibility that infants perceived only the initial tone as opposed to the tonal configuration. From the studies to be reported here, it should become clear that, at the very least, infants can discriminate contrasting tonal sequences.

Relations within Patterns

Melodic Transformations

Consider, for the moment, a pattern comprised of several tones of different frequencies. When such a sequence is shifted or transposed to different absolute frequencies, so long as the components remain invariant in relation to one another, the shifted pattern is perceived as equivalent to the original pattern (Dowling & Fujitani, 1971). In musical terms, we speak of recognizing the melody despite a change in key. This tendency to code relational information in a tune is so strong that it appears to be easier to remember the relational information than to retain the absolute pitch of component tones (Deutsch, 1969, 1975; but see Terhardt & Ward, 1982). With familiar or highly overlearned material, we remember rather precise information about the pitch relations or intervals between tones. With less familiar material, we appear to lose some of this precision, but we still remember the overall direction of pitch changes or the melodic contour (Attneave & Olson, 1971; Bartlett & Dowling, 1980; Dowling, 1978).

Chang and I (Chang & Trehub, 1977a) attempted to determine whether 5-month-old infants could recognize transpositions of 6-tone sequences. Using the habituation–dishabituation design, we presented a sequence of tones repeatedly, then presented half of the infants with a transposition (3 semitones higher or lower) that altered the frequencies of component tones but preserved the exact frequency ratios between tones. The remaining infants heard a pattern with the component tones of the transposition reordered. Since both groups of infants received patterns comprised of identical elements, any difference in response would have to be attributed to the presence or absence of relational information. The results revealed significant dishabituation or response recovery to the pattern that failed to preserve the original relations, indicating that infants detected the change. In contrast, there was no response recovery to the transposed pattern, suggesting that infants "recognized" the familiar melodic pattern under conditions of transposition (i.e., the frequency relations between tones). Chang and I obtained fairly similar results when we subsequently tested 2-month-old infants. From these findings there was no basis for determining whether infants were coding precise interval information or whether they were simply tracking the general contour (i.e., sequence of ups and downs) of the patterns. Nevertheless, Dowling (1982) and Pick (1979) have suggested that the global nature of contour information compared to the greater precision of interval information should predict the earlier application of a contour-processing strategy.

Recently, we have been addressing this issue with 8- to 11-month-old infants (Trehub *et al.*, 1984). The habituation–dishabituation task in the previous study (Chang & Trehub, 1977a) had been a boring one for infants, with test trials separated by 15 sec, and many infants choosing to fuss, cry, or fall asleep rather than complete the procedure. With our alternative procedure, infants began by listening to repeating 6-tone patterns. (The tones were sinusoidal waveforms, 200 msec in duration, with intertone pauses of 200 msec and interpattern intervals of 800 msec.) They were trained to turn whenever a change occurred in the contour and frequency range of a melody, then tested on equal numbers of change and no-change (control) trials. On change trials, infants were presented with one of several types of transformations including those that preserved contour and interval information (i.e., transpositions), those that preserved contour but not exact intervals, and those that violated contour. Infants discriminated all transformations from the standard and their performance on transpositions and contour-preserving melodies did not differ. Thus, with very brief retention intervals, there is evidence that memory for the absolute pitch of component tones remains intact. These results are consistent with Bartlett and Dowling's (1980) finding that, with short retention intervals, adults' detection of key similarity is much easier than their detection of the invariance of melodic intervals. They speculate that the relative importance of key information in memory decreases drastically as the retention interval is lengthened, whereas the relative importance of interval information increases.

In a subsequent experiment (Trehub *et al.*, 1984, Experiment 2) task difficulty was increased by inserting a distractor sequence (three identical tones, each of 200-msec duration with 200-msec intertone intervals) between repetitions of the standard melody and between the standard and transformed melody. Infants were then tested with exact transpositions, contour-preserving transformations, and contour-violating transformations. Under these conditions of increased difficulty, infants showed no evidence of discriminating the transpositions or the contour-preserving transformations from the standard melody. They did, however, detect the contour-violating transformations, including those that comprised a reordering of identical tones. The implication here is that infants treated transposed and contour-preserving melodies as similar or equivalent to the original melody and that they did so by encoding and retaining information about contour as opposed to absolute intervals. This is consistent with findings of adults' retention of the contour rather than the absolute intervals of atonal melodies or unfamiliar tonal melodies (Dowling, 1978, 1982; Dowling & Fujitani, 1971). In short, then, infants' perception of melodies appears to be holistic or structured, with the global property of contour perceived across transformations of specific properties such as absolute frequency and interval size.

Temporal Grouping or Rhythm

Not only do we structure frequency relations in tone sequences, but we also structure temporal information concerning the duration of tones and intertone intervals. As a result, we organize sequences of tones into subgroups or small configurations within a larger configuration. We create these temporal groupings only partially from duration and interval information. In many cases, the perceived groupings may not conform to the actual physical spacing of stimuli, being influenced by other physical attributes of the stimulus, such as the frequency or intensity relations between tones.

Psychologists have been aware of these rhythmic phenomena for quite some time. As early as 1894, it was reported that a loud sound altered the perceived interval between preceding and following sounds (Meumann, 1894). Several researchers have also reported that the intervals between conceptual groupings appear longest (Bolton, 1894; Fitzgibbons, Pollatsek, & Thomas, 1974; MacDougall, 1903; Woodrow, 1909). Indeed, the tendency to impose rhythmic structure is so compelling that even successive clicks or tones that are uniform in all respects (frequency, intensity, and duration) tend to be grouped, with the first element of each group gradually appearing to be accentuated (Bolton, 1894; Fraisse, 1982; Miner, 1903; Seashore, 1938; Woodrow, 1909). Temporal grouping has implications beyond phenomenal changes in perception, for such grouping, whether subjectively or objectively based, has considerable influence on our memory for patterns (Aaronson, 1967; Huttenlocher & Burke, 1976; Ryan 1969; Winzenz & Bower, 1970).

With respect to infants, Chang and I (Chang & Trehub, 1977b) found that 5-month-olds could distinguish between 6-tone sequences with identical tones but contrasting temporal groupings. Demany, McKenzie, and Vurpillot (1977) also found discrimination of rhythmic changes but with different rhythms and with younger infants (2- to 3-months-old). These findings, although confirming that rhythmic changes are perceptible, convey little about infants' organizational propensities in this domain and nothing at all about the issue of subjective grouping.

The issue is not well understood, even with adults. In most experiments, listeners have been required to convey information about their groupings either by verbal description or by tapping out heard rhythms. The extent to which such production tasks reflect perceptual processes is unclear.

To gain further insight into the phenomenon of subjective grouping, we have done some preliminary work with adults (Trehub, Thorpe, & Bull, in preparation). We have been using the following auditory pattern: three tones of identical frequency (440 or 659 Hz) followed by three tones of higher or lower frequency (659 or 440 Hz), with the duration of all tones and intertone intervals identical (200 msec). This pattern of frequencies provides a compelling organization: presumably, two groups of three tones. Does this mean that the interval or gap between groups is *heard* as being longer than the intervals or gaps within groups?

We attempted to specify the perceived duration of the between-group interval by comparing it with actual intervals of varying magnitude. We did this with an *AX* design in which listeners judged whether the second pattern (*X*) was the same as the first (*A*). *A* was the pattern with equivalent intertone intervals and *X* had one of the intertone intervals lengthened by 0, 20, 40, 60, or 80 msec. Our prediction was that it would be easier to perceive a change that violated the hypothesized grouping by frequency than one that was consonant with that organization; that is, changes in any noncentral interval would be more readily detected than changes in the central interval. Our predictions were realized. When one interval of the pattern was extended, detection of the change was clearly inferior at the central location compared to any other location.

We have also investigated this question with 6- to 8-month-old infants (Trehub *et al.*, in preparation) using the operant head-turn procedure, as in our melodic transformation studies (Trehub *et al.*, 1984). We presented repetitions of the standard melody used with adults (three 659-Hz tones followed by three 440-Hz tones) and trained infants to turn when the interval or gap following the second tone was extended by 400 msec. Following training, independent groups of infants received change trials with an interval extended by 200, 125, and 100 msec following either the third or fourth tone; they also received an equivalent number of no-change trials. We predicted that infants would detect the extended gap within a frequency grouping (i.e., following the fourth tone) more readily than that between frequency groupings (i.e., following the third tone). In this case, our predictions were not confirmed. Infants detected all changes from the standard melody, performing equivalently for both gap locations at each value tested. There was, however, an orderly decrement in performance as the size of the increment decreased. When the gap increment was reduced further to 75 msec there was some indication of a grouping effect. In this case, only the gap at the noncentral interval was detected.

In a subsequent experiment (Thorpe, Trehub, & Morrongiello, in preparation), we increased the frequency separation between low and high tones (from 659 and 440 Hz to 1600 and 160 Hz, respectively) and found that infants' detection of the extended central gap was significantly poorer than their detection of the extended noncentral gap. In fact, infants were unable to detect a gap increment of 100 msec at the central location even though they could do so with smaller frequency differences. These findings are consistent with evidence that adults' perception of gaps between successive tones becomes less accurate as the frequency separation between tones is increased (Divenyi & Danner, 1977; Divenyi & Sachs, 1978; Williams & Perrott, 1972).

Stream Segregation or Temporal Regrouping

Although rhythmic grouping affects the perceived distribution of stimulus elements over time, it preserves the physical ordering of elements. In contrast, the phenomenon of stream segregation (Bregman, 1978a,b,c, 1981; Bregman & Campbell, 1971; Bregman & Dannenbring, 1973; Bregman & Rudnicky, 1975; Bregman & Steiger, 1980; Dannenbring & Bregman, 1976, 1978) brings about the decomposition and reordering of elements. Repeated sequences of discrete sounds that are presented at rapid rates tend to "split" perceptually into two or more distinct but overlapping sequences. Each sequence is comprised of elements that are similar in some respect, such as frequency, overtone structure, or duration. Listeners can track the order of elements within a stream but not across streams. On the surface, the phenomenon seems counterproductive in its "destruction" of the presented configuration. Bregman (1978c) notes, however, that the grouping of elements on the basis of commonalities promotes coherence with real world stimuli if not those of the laboratory. Recently, Demany (1982) has found that such segregation of streams by frequency similarity occurs for infants as young as 7 weeks. Like adults, they detect changes in the order of elements of a rapidly presented sequence only when the frequency differences are relatively small. We are currently attempting to ascertain the principles underlying the segregation of auditory streams by infants and young children.

Relations between Patterns

Perceptual Constancy

Holistic or relational processing is also involved in our ability to extract consistent phonemic information from different speakers, speaking rates, and contexts. Kuhl and her associates (Kuhl, 1979, 1980, 1983; Kuhl & Miller, 1982) attempted to investigate this phenomenon with infants, and accumulated evidence of discrimination as opposed to actual categorization. The essence of Kuhl's approach was to determine whether infants could maintain discrimination between various speech segments in the presence of irrelevant and potentially distracting changes in voice quality or pitch contour. Evidence for such discrimination was interpreted as indicating infants' perception of the similarity of varying exemplars and, by implication, their ability to organize stimuli according to the acoustic properties that underlie phonetic categories.

Following on the heels of Kuhl's insights, we have been investigating somewhat comparable questions with nonspeech stimuli (Endman, 1984). Essentially, we have been examining the ability of 7- to 8-month-old infants to attend to differences in the harmonic structure of nonspeech patterns while ignoring variations in fundamental frequency, intensity, or duration. Our stimuli are speechlike envelopes and the task is the operant head-turning procedure described earlier. The infant is presented with repetitions of a set of sounds with a particular harmonic structure (e.g., energy peaks at 570 and 840 Hz), the exemplars embodying four variations of fundamental frequency, intensity, or duration, presented in random order. During brief test trials (4 sec), a similar set of sounds with contrasting harmonic structure (e.g., energy peaks at 270 and 2300 Hz) is presented with the aforementioned variations. This can be regarded as a primitive concept formation task, with two levels of the relevant dimension (harmonic structure) and four levels of the irrelevant dimension (fundamental frequency, intensity, or duration). The infant must demonstrate the use of a "rule" for selecting positive and negative instances (i.e., for turning to changes in harmonic structure and not turning to changes in fundamental frequency, intensity, or duration).

The task turns out to be very easy for infants. Despite the fact that infants were not presented with variations in the irrelevant dimension during the training period, they continued to respond to the relevant change (in this case, a specific harmonic structure) when the variable exemplars were introduced. It could be argued that infants simply learned the identity of the four standard stimuli (i.e., four variations of one harmonic structure) and thus could readily ascertain that the change stimuli had not been heard previously. This has been ruled out by one condition in which the two sets were composed randomly from the eight stimuli; infants showed no evidence of discriminating these sets.

Aslin, Pisoni, and Jusczyk (1983) have claimed that infants' discrimination of arbitrary groupings of the stimuli used in Kuhl's studies would provide some measure of confidence that they recognized the similarity between varying exemplars from the same phonetic category. Accordingly, infants in the Endman (1984) study can be said to have met the Aslin *et al.* (1983) criterion for perceptual constancy. It could be argued, further, that infants succeeded on the Endman *et al.* task because they capitalized on the special qualities that distinguish speech categories (e.g., Eimas, 1975; Eimas & Miller, 1980).

On the other hand, it might be the case that infants could be trained to respond, in comparable fashion, to changes in other acoustic dimensions that do not map onto phonetic categories. Success on such a task would argue for greater generality of the phenomenon in question. In an attempt to address this question, we reversed the relevant and irrelevant dimensions of the previous experiment (Endman, 1984). The target or the to-be-discriminated dimension was now fundamental frequency, intensity, or duration for three independent groups of infants. Correspondingly, the irrelevant dimension was harmonic structure, with four variations in each case. As before, infants performed all discriminations with ease. These findings imply that there is considerable flexibility in infants' ability to group stimuli on the basis of similarity. When they are rewarded for focusing on parameters normally associated with phonetic differences and ignoring other discernible differences, they do so. When the payoff is different, however, so is the pattern of responding. The limits of such flexibility remain to be determined.

Concluding Comments

The problem posed by adults' organization of sensory information is no less an issue today (e.g., Hochberg, 1979; Kubovy & Pomerantz, 1981) than it was 100 years ago (Boring, 1942; Helmholtz, 1877/1954). Recognizing a melody over transformations of its component notes, a word over variations in vocal quality, and a visual object over retinal changes are commonplace phenomena that continue to elude simple explanation. In the specific realm of music perception, one strategy has been to compare the pattern analyzing abilities of experienced and inexperienced listeners (i.e., skilled musicians and nonmusicians). With respect to music as well as other perceptual domains, infants represent an extreme case of inexperience. For this reason alone, their study holds promise for important revelations.

References

Aaronson, D. Temporal factors in perception and short-term memory. *Psychological Bulletin,* 1967, *67,* 130–144.

Aslin, R. N., Pisoni, D. B., & Jusczyk, P. W. Auditory development and speech perception in infancy. In P. Mussen & M. Hetherington (Eds.), *Handbook of child psychology (Vol 2): Perceptual Development.* New York: Wiley, 1983.

Attneave, F. Some informational aspects of visual perception. *Psychological Review,* 1954, *61,* 183–193.

Attneave, F., & Olson, R. K. Pitch as a medium: A new approach to psychophysical scaling. *American Journal of Psychology,* 1971, *84,* 147–166.

Bartlett, J. C., & Dowling, W. J. Recognition of transposed melodies: A key-distance effect in developmental perspective. *Journal of Experimental Psychology: Human Perception and Performance,* 1980, *6,* 501–515.

Bolton, T. L. Rhythm. *American Journal of Psychology,* 1894, *6,* 145–238.

Boring, E. G. *Sensation and perception in the history of experimental psychology.* New York: Appleton-Century-Crofts, 1942.

Bregman, A. S. Auditory streaming: Competition among alternative organizations. *Perception and Psychophysics,* 1978, *23,* 391–398. (a)

Bregman, A. S. Auditory streaming is cumulative. *Journal of Experimental Psychology: Human Perception and Performance,* 1978, *4,* 380–387. (b)

Bregman, A. S. The formation of auditory streams. In J. Requin (Ed.), *Attention and performance* (Vol. VII). Hillsdale, N.J.: Erlbaum, 1978. (c)

Bregman, A. S. Asking the "what for" question in auditory perception. In M. Kubovy & J. R. Pomerantz (Eds.), *Perceptual organization.* Hillsdale, N.J.: Erlbaum, 1981.

Bregman, A. S., & Campbell, J. Primary auditory stream segregation and perception of order in rapid sequences of tones. *Journal of Experimental Psychology,* 1971, *89,* 244–249.

Bregman, A. S., & Dannenbring, G. L. The effect of continuity on auditory stream segregation. *Perception and Psychophysics,* 1973, *13,* 308–312.

Bregman, A. S., & Dannenbring, G. L. Auditory continuity and amplitude edges. *Canadian Journal of Psychology,* 1977, *31,* 151–159.

Bregman, A. S., & Rudnicky, A. I. Auditory segregation: Stream or streams? *Journal of Experimental Psychology: Human Perception and Performance,* 1975, *1,* 263–267.

Bregman, A. S., & Steiger, H. Auditory streaming and vertical localization: Interdependence of "what" and "where" in decisions in audition. *Perception and Psychophysics,* 1980, *80,* 539–546.

Broadbent, D. E., & Ladefoged, P. Auditory perception of temporal order. *Journal of the Acoustical Society of America,* 1959, *31,* 1539.

Chang, H. W., & Trehub, S. E. Auditory processing of relational information by young infants. *Journal of Experimental Child Psychology,* 1977, *24,* 324–331. (a)

Chang, H. W., & Trehub, S. E. Infants' perception of temporal grouping in auditory patterns. *Child Development,* 1977, *48,* 1666–1670. (b)

Dannenbring, G. L., & Bregman, A. S. Stream segregation and the illusion of overlap. *Journal of Experimental Psychology: Human Perception and Performance,* 1976, *2,* 544–555.

Dannenbring, G. L., & Bregman, A. S. Streaming vs. fusion of sinusoidal components of complex tones. *Perception and Psychophysics,* 1978, *24,* 369–376.

Demany, L. Auditory stream segregation in infancy. *Infant Behavior and Development,* 1982, *5,* 261–276.

Demany, L., McKenzie, B., & Vurpillot, E. Rhythm perception in early infancy. *Nature,* 1977, *266,* 718–719.

Deutsch, D. Music recognition. *Psychological Review,* 1969, *76,* 300–307.

Deutsch, D. The organization of short-term memory for a single acoustic attribute. In D. Deutsch & J. A. Deutsch (Eds.), *Short-term memory.* New York: Academic Press, 1975.

Divenyi, P. L., & Danner, W. F. Discrimination of time intervals marked by brief acoustic pulses of various intensities and spectra. *Perception and Psychophysics,* 1977, *21,* 125–142.

Divenyi, P. L., & Sachs, R. M. Discrimination of time intervals bounded by tone bursts. *Perception and Psychophysics,* 1978, *24,* 429–436.

Dowling, W. J. Scale and contour: Two components of a theory of memory for melodies. *Psychological Review,* 1978, *85,* 341–354.

Dowling, W. J. Melodic information processing and its development. In D. Deutsch (Ed.), *The psychology of music.* New York: Academic Press, 1982.

Dowling, W. J., & Fujitani, D. A. Contour, interval, and pitch recognition in memory for melodies. *Journal of the Acoustical Society of America,* 1971, *49,* 524–531.

Eilers, R. E., Wilson, W. R., & Moore, J. M. Developmental changes in speech discrimination in infants. *Journal of Speech and Hearing Research,* 1977, *20,* 766–780.

Eimas, P. D. Auditory and phonetic coding of the cues for speech: Discrimination of the [*r*-l] distinction by young infants. *Perception and Psychophysics,* 1975, *18,* 341–347.

Eimas, P. D., & Miller, J. Contextual effects in infant speech perception. *Science,* 1980, *209,* 1140–1141.

Endman, M. *Perceptual constancy for nonspeech stimuli.* Paper presented at the International Conference on Infant Studies, New York, April, 1984.

Fitzgibbons, P. J., Pollatsek, A., & Thomas, I. B. Detection of temporal gaps within and between groups. *Perception and Psychophysics,* 1974, *16,* 522–528.

Foulke, E. & Sticht, T. G. Review of research in the intelligibility and comprehension of accelerated speech. *Psychological Bulletin,* 1969, *72,* 50–62.

Fraisse, P. Rhythm and tempo. In D. Deutsch (Ed.), *The psychology of music.* New York: Academic Press, 1982.

Garner, W. R. *Uncertainty and structure as psychological concepts.* New York: Wiley, 1962.

Garner, W. R. *The processing of information and structure.* Potomac, Maryland: Erlbaum, 1974.

Helmholtz, H. von. *On the sensations of tone as a physiological basis for the theory of music.* (A. J. Ellis, Ed. and trans.) New York: Dover, 1954. (Originally published in German, 1877.)

Hochberg, J. Sensation and perception. In E. Hearst (Ed.), *The first century of experimental psychology.* Hillsdale, N.J.: Lawrence Erlbaum Associates, 1979.

Huttenlocher, J., & Burke, D. Why does memory span increase with age? *Cognitive Psychology,* 1976, *8,* 1–31.

Jones, M. R. Auditory patterns: The perceiving organism. In E. C. Carterette & M. P. Friedman (Eds.), *Handbook of perception* (Vol. 8). New York: Academic Press, 1978.

Kinney, D. K. & Kagan, J. Infant attention to auditory discrepancy. *Child Development,* 1976, *47,* 155–164.

Koffka, K. *The principles of Gestalt psychology.* New York: Harcourt Brace, 1935.

Kubovy, M. & Pomerantz, J. R. (Eds.). *Perceptual organization.* Hillsdale, N.J.: Erlbaum, 1981.

Kuhl, P. K. Speech perception in early infancy: Perceptual constancy for spectrally dissimilar vowel categories. *Journal of the Acoustical Society of America,* 1979, *66,* 1668–1679.

Kuhl, P. K. Perceptual constancy for speech-sound categories. In G. H. Yeni-Komshian, J. F. Kavanaugh, & C. A. Ferguson (Eds.), *Child phonology (Vol. 2): Perception.* New York: Academic Press, 1980.

Kuhl, P. K. The perception of speech in early infancy: Four phenomena. In S. E. Gerber & G. T. Mencher (Eds.), *The development of auditory behavior.* New York: Grune & Stratton, 1983.

Kuhl, P. K., & Miller, J. D. Discrimination of auditory target dimensions in the presence of variation in a second dimension by infants. *Perception and Psychophysics,* 1982, *31,* 279–292.

Leavitt, L. A., Brown, J. W., Morse, P. A., & Graham, F. K. Cardiac orienting and auditory discrimination in 6-week infants. *Developmental Psychology,* 1976, *12,* 514–523.

MacDougall, R. The structure of simple rhythm forms. *Psychological Review, Monograph Supplements,* 1903, *4,* 309–416.

McCall, R. B., & Melson, W. H. Amount of short-term familiarization and the response to auditory discrepancies. *Child Development,* 1970, *41,* 861–869.

Melson, W. H., & McCall, R. B. Attentional responses of five-month-old girls to discrepant auditory stimuli. *Child Development,* 1970, *41,* 1159–1171.

Meumann, E. Untersuchungen zur Psychologie und Aesthetik des Rhythmus. *Philosophische Studien,* 1894, *10,* 249–322, 393–430.

Miner, J. B. Motor, visual, and applied rhythms: An experimental study and a revised explanation. *Psychological Review, Monograph Supplements,* 1903, *3,* No. 4 (whole No. 21).

Neisser, U. *Cognitive psychology.* New York: Appleton-Century-Crofts, 1967.

Pick, A. D. Listening to melodies: Perceiving events. In A. D. Pick (Ed.), *Perception and its development.* New York: Wiley, 1979.

Pomerantz, J. R. Perceptual organization in information processing. In M. Kubovy & J. R. Pomerantz (Eds.), *Perceptual organization.* Hillsdale, N.J.: Erlbaum, 1981.

Ryan, J. Temporal grouping, rehearsal and short-term memory. *Quarterly Journal of Experimental Psychology,* 1969, *21,* 148–155.

Seashore, C. E. *Psychology of music.* New York: McGraw-Hill, 1938.

Terhardt, E., & Ward, W. D. Recognition of musical key: Exploratory study. *Journal of the Acoustical Society of America,* 1982, *72,* 26–33.

Thorpe, L. A., Trehub, S. E., & Morrongiello, B. A. *Infants' perception of melodies: Further observations on temporal grouping.* Manuscript in preparation.

Trehub, S. E. Infants' sensitivity to vowel and tonal contrasts. *Developmental Psychology,* 1973, *9,* 81–96.

Trehub, S. E. The discrimination of foreign speech contrasts by infants and adults. *Child Development,* 1976, *47,* 466–472.

Trehub, S. E., Bull, D., & Thorpe, L. A. Infants' perception of melodies: The role of melodic contour. *Child Development,* 1984, *55,* 821–830.

Trehub, S. E., Thorpe, L. A., & Bull, D. *Infants' perception of melodies: The role of temporal grouping.* Manuscript in preparation.

Warren, R. M., Obusek, C. J., & Ackroff, J. M. Auditory induction: Perceptual synthesis of absent sounds. *Science,* 1972, *176,* 1149–1151.

Warren, R. M., Obusek, C. J., Farmer, R. M., & Warren, R. P. Auditory sequence: Confusions of patterns other than speech or music. *Science,* 1969, *164,* 586–587.

Williams, K. N., & Perrott, D. R. Temporal resolution of tonal impulses. *Journal of the Acoustical Society of America,* 1972, *51,* 644–647.

Winckel, F. *Music, sound, and sensation.* New York: Dover, 1967.

Winzenz, D., & Bower, G. H. Subject-imposed coding and memory for digit series. *Journal of Experimental Psychology,* 1970, *83,* 52–56.

Woodrow, H. A quantitative study of rhythm: The effect of variations in intensity, rate and duration. *Archives of Psychology,* 1909, *18,* No. 1.

CHAPTER 11

Infant Speech Perception

Environmental Contributions

Rebecca E. Eilers and D. Kimbrough Oller

Mailman Center for Child Development
Departments of Pediatrics and Psychology
University of Miami
P.O. Box 016820
Miami, Florida

Backdrop

The formal study of infant speech perception was inaugurated by Eimas, Siqueland, Jusczyk, and Vigorito (1971), who assumed a radical theoretical position in asserting that infant perceptual abilities with regard to speech are "linguistic" and "innate." Since then, research on the speech perceptual capabilities of infants has expanded enormously, and a major controversy has developed around the nature–nurture question. In order to introduce the evidence feeding this controversy, as well as the conclusions based on this evidence, the following listing of key empirical results and interpretations is provided. The studies mentioned are representative of the field in general and illustrate the theoretical confusion that has characterized this domain.

1. Eimas *et al.* (1971) found that 1- to 4-month-old infants could discriminate between two synthetic English syllables, /ba/ and /pha/, differing in a feature called "voicing".[1] Eimas and colleagues concluded that speech perception is innate and is accomplished in a linguistic mode.

[1]Voicing of a consonant is determined by the timing of vocal fold vibration with respect to the opening articulatory gesture toward an adjacent vowel. In English /ba/, vibration of vocal folds usually begins shortly after (less than 30 msec) the lips open toward the vowel /a/. In /pha/ the vibration begins later (over 30 msec).

2. Trehub and Rabinovitch (1972) found that infants could discriminate natural as well as synthetically produced English voicing contrasts /ba/ and /pha/ and concluded that infants can detect differences in voicing. But these authors expressed reluctance to interpret the results as indicating special linguistic abilities in infants.

3. Morse (1972) and later Till (1976) found that infants could discriminate contrasts in place of articulation (/ba/ versus /ga/) and concluded that the infant's ability to discriminate rapidly changing spectral information was innate and that infants respond to the acoustic cues for place of articulation in a "linguistically relevant manner."

4. Moffitt (1971) found that 4-month-old infants were able to discriminate among syllables differing in place of articulation and concluded that "linguistic perceptual capacities are present during early life."

5. Eilers and Minifie (1975) found that while infants could discriminate among certain fricative consonants, for example, /ʃa/ versus /sa/ and/va/ versus /sa/, 3- to 4-month-old infants failed to provide evidence of discriminating fricative consonants cued solely by voicing (/sa/ versus /za/). They concluded that not all speech contrasts were equally discriminable and that a continuum of difficulty existed.

6. Butterfield and Cairns (1974), reviewing the work of Eimas *et al.* (1971) and Moffitt (1971), suggested that while English-learning infants may discriminate English voicing contrasts, they do not demonstrate discrimination of Spanish-like voicing contrasts. Butterfield and Cairns concluded that infants' innate abilities may be neither linguistic nor general.

7. Lasky, Syrdal-Lasky, and Klein (1975) and Streeter (1976) found that infants learning Spanish and Kikuyu (an African language), respectively, were able to discriminate the Spanish and Kikuyu-like consonant contrast of voicing, a contrast quite different from that occurring in English. They concluded that since American-English children seemed not to discriminate the Spanish/Kikuyu contrast (based on earlier work reviewed by Butterfield and Cairns, 1974), the achievement of the Spanish/Kikuyu babies represented an interaction of environment and innate abilities.

8. Eilers, Gavin, and Wilson (1979) found that Spanish-learning infants performed significantly better than English-learning infants on Spanish-like voicing contrasts, while both groups discriminated the English voicing contrast. Eilers, Gavin, and Oller (1982) also found that Spanish-learning infants performed significantly better on the Spanish tap-trill[2] distinction than English-learning children. They concluded that early experience influences phonemic

[2]Spanish contrasts a tapped "r" similar to the medial element in bu*tt*er with a trilled "r," which includes repetitive tapping. The contrast presented to infants consisted of a single tap versus a trill with two taps.

perception and that the Spanish-like voicing contrast is inherently (auditorily) more difficult than the English voicing contrast.

9. Eilers, Morse, Gavin, and Oller (1981) found that infants were unable to provide evidence of discrimination of English voicing contrasts differing only in a single aspect of voicing, the onset time of voicing of the first formant. The same infants easily discriminated voicing contrasts with multiple acoustic cues, such as had been used in the earlier Eimas *et al.* (1971) studies. They concluded that the Eimas *et al.* (1971) results may have been misinterpreted. Infants in the earlier study may have been responding to a conjunct of acoustic cues and not to a single linguistically relevant cue.

10. Eilers, Wilson, and Moore (1977) found that 12-month-old infants perform more accurately on a variety of speech perception tasks than 6-month-olds.

11. Tallal and Stark (1980) reported that perceptual constancy tasks similar to those demonstrated with 6-month-olds by Kuhl (1976) are difficult to show in 2- to 3-year-olds. Oller and Eilers (1983) reported difficulties in identification by English and Spanish 2-year-olds of nonnative contrasts presented in a real speech setting. This same difficulty was not evident with native contrasts presented in the same made-up speech game.

12. Miyawaki, Strange, Verbrugge, Liberman, Jenkins, and Fujimura (1975) showed failure of adults to categorically discriminate foreign distinctions that English-learning babies discriminate (/r/ and /l/). In contrast, Eilers, Wilson, and Moore (1979) and Carney, Widin, and Viemeister (1977) showed that adult capabilities may be sharper than those of infants, by demonstrating that adults can discriminate within category as well as across category boundaries.

An Interpretive Model

Assuming that all of these data are accurate, it is clear that a model to account for them must be more complex than that posited in the earliest infant speech perception studies. Findings that indicate some development of skills across the first years of life, differences in linguistic-perceptual abilities between groups of infants whose language backgrounds differ, *poor* performance by adults on foreign contrasts in some conditions, as well as *good* performance by adults on foreign contrasts under other test conditions point to some role for environment as well as complex innate auditory abilities.

On the basis of our own work (Eilers & Minifie, 1975; Eilers, Wilson, & Moore, 1977; Eilers, Gavin, & Wilson, 1979; Eilers & Gavin, 1981; Eilers, Gavin, & Oller, 1982), we have posited a multifaceted model that includes a role for linguistic experience as well as innate predispositions. In this model, innate features include complex auditory skills but not specifically *linguistic* processing capabilities, since it appears that existing data can be accounted for without reference to linguistic processing.

The present paper adds to the theoretical description by presenting further features of a model with explanatory and predictive capabilities. The goal of the theory is to present a more integrated view than previously available, including clearer hypotheses concerning the development of speech-processing abilities and a more detailed description of the course of development.

Since we assume that the child's basic auditory-processing capabilities may be common to mammalian species, we need to ask how the child becomes more linguistically oriented. The characterization of what it means for the child to become more linguistic in his speech processing requires a model of how a linguistic speech processor might operate. Let us, then, consider certain necessary levels of analysis through which a mature listener presumably codes acoustic speech information on the way to determining meaning. The present discussion will outline a four-level acoustic-to-phonemic processor. Actual processors may require additional levels between acoustic and phonemic levels (e.g., phrase-level or word-level breakdowns of acoustic data) and will clearly require a number of levels beyond the phonemic. The present discussion will be devoted to syllabic and segmental processing only.

In processing a speech signal (see Figure 1), one of the early steps requires that the acoustic stream be broken into small analyzable chunks based on a procedure that locates syllables, or at least syllabic prominences in the waveform. The output of the procedure would be a phonetically rudimentary syllabic sequence, available for future analysis into phonetic segments. In a later step, the syllables must be analyzed for specific phonetic units, some of which are primarily associated with prominences in the power envelope (vowels) and some

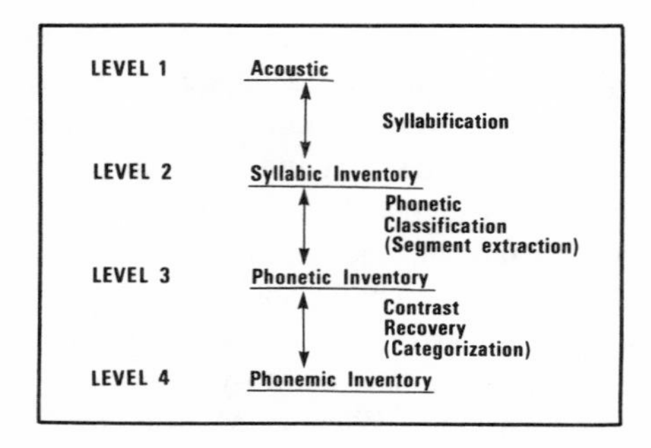

Figure 1. Model of infant speech perception: basic acoustic-to-phonemic processing levels.

primarily associated with low-points in the envelope (consonants). The output of this second procedure would be a set of phonetically described segmental elements including substantial allophonic (featural) and prosodic detail, available for future categorization in terms of the broader phonemic units of the speech code. A still later procedural step is to determine the phonemic character of the phonetically specified elements. The output of this step provides contrastive, featurally specified, relatively abstract units. Speech comprehension then proceeds through several additional processing levels toward recovery of meaningful units.

Although a particular chunk of speech information can be processed in the fashion outlined, it is not necessarily the case (especially in long utterances) that all of a sequence is analyzed at one level before the processing of another level begins. More likely, smaller chunks are handled in parallel, or at least small chunks (analyzed out by utterance, phrase or word-level preprocessors) are processed while others are being held in short-term storage buffers.

It is also important to point out that although there are instances in which large-chunk processing events logically *must* proceed in the order given, there are clearly other instances in which certain steps are bypassed based on predictability of the message. For example, if I say "rheumatic fe——" there is reason to predict that the next syllable will be "ver." Although a human listener may in such a case verify cursorily that the expected syllable does occur, it is usually unnecessary to expend the same level of processing effort to determine the syllable's nature as in other less predictable cases. Warren and Obusek (1971) have illustrated the predictive processing of human speech perceivers by revealing perceptual illusions associated with phonetic prediction. Our model does not, then, require that adult listeners always recover syllabic, phonetic, or phonemic information about each acoustic chunk, since often that information can be predicted on the basis of previously determined information. The model merely claims that in cases where predictability is low, the steps of processing necessary to recover information efficiently must include those diagrammed in Figure 1.

Considering our model in terms of empirical data about speech discrimination in the first months of life, it is unlikely that the infant codes information in terms of any of the posited units below the acoustic level. Results of studies of infant discrimination appear to be explainable in terms of basic auditory abilities without including specifically linguistic syllabification, phonetic classification, or contrast recovery. This is not to say that the auditory processing of the infant is simple, or to say necessarily that the output of the infant's processor is not an abstraction of the acoustic signal. In fact, the infant may bring a variety of general cognitive strategies for processing sensory information to the task of auditory analysis. The infant's coding, however, is not necessarily speech-specific.

If it is correct that the infant does not initially process acoustic information in the way mature linguistic processors do, it may be that an early and major phonological achievement is the development of a syllabification procedure incorporating information about mature syllable timing and contours. Although the infant's innate auditory capacities may include global strategies for temporarily chunking information, the process of syllabification may include more precise specification of timing parameters for various kinds of speech events. The result of the application of syllabification would be that the infant would have at his disposal a second level of processing in addition to the basic acoustic level (see Figure 2).

What does the child achieve by coding acoustic information syllabically? The child simplifies the task of processing speech by chunking information appropriately and providing a basis for efficiently locating the acoustic segments to be analyzed in special detail. Syllabification provides the time frame that tells the child where to look for specific features and it may be a process necessary for adequate treatment of multisyllabic speech.

In a subsequent stage of development, according to the model, the child further breaks apart the levels of representation by analyzing syllabic information into specific phonetic units. These units would be abstract in that they represent groupings of acoustic-syllabic patterns into hierarchically defined categories. For example, syllabic units such as [sa], [as], [si], [is], [su], [us] might be reanalyzed to indicate that all include a common consonantal element [s], which had not previously been treated as a unit. Further, phonetic-feature categories may attain fuller status (e.g., [s] and [z] might both be noted to be alveolar fricatives). The procedure for phonetic grouping would be more efficient and would achieve greater generality due to the prior syllabic coding. Figure 3 diagrams the child's system at Stage 3.

What is the advantage of the phonetic coding? The inventory of different syllabic types includes a considerable array of acoustic patterns. By coding these phonetically, the infant or child analyzes the components of many syllable types

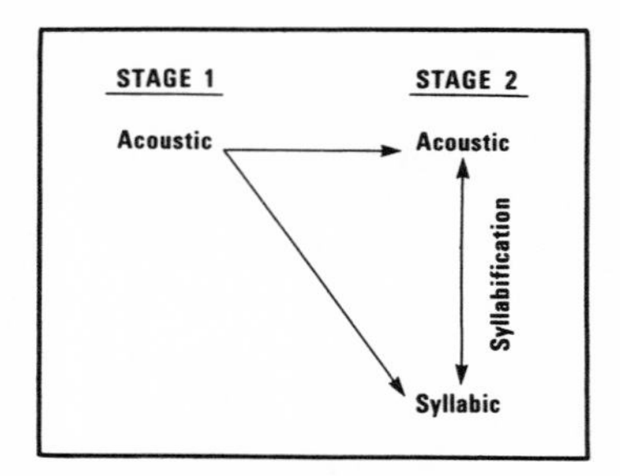

Figure 2. Model of infant speech perception: addition of syllabic level.

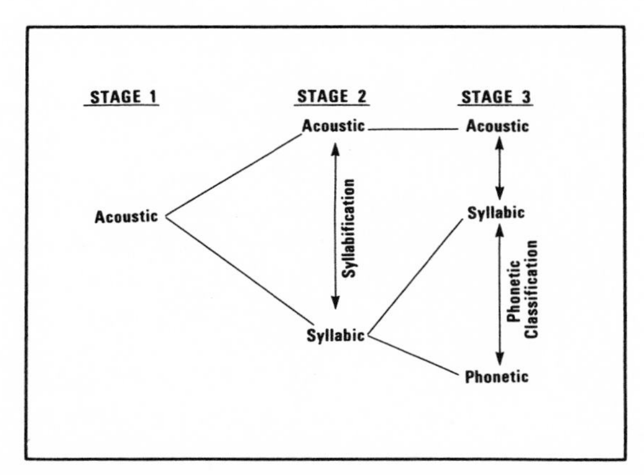

Figure 3. Model of infant speech perception: addition ot phonetic level.

and sorts them into broader and more manageable sets of units for further processing.[3]

In the fourth proposed stage, the infant comes to recode the phonetic-contrast types. The separation of phonemic levels involves formulation of a set of rules of allophonic variation, that is, rules specifying the various phonetic forms that a phonemic type may assume depending on context (e.g., the abstract phonemic element /b/ would be specified as having prevoiced or unvoiced unaspirated phonetic variants depending on position and stress). The phonemic level of representation is added to the child's system in Stage 4 (see Figure 4).

What does the child achieve by contrast organization? The phonetic elements to be manipulated are recoded for organizational simplicity. The list of possible phonetic elements is relatively long and cumbersome. The phonemic recoding of these reduces the set to a small, hierarchically arranged inventory specific to the child's native language.

Each of the steps of development posited here represents an achievement that makes the infant's or child's processing more efficient and more capable of handling broader and more complex inventories. At the beginning of speech discrimination, the infant may have treated such small inventories of acoustic types, that an ACOUSTIC-only strategy for processing was adequate. As more complex tasks are addressed, with inventories increasing in size, more efficient approaches are necessary and ''linguisticization'' of the processing approach becomes advantageous.

[3]To say that the infant at this stage codes syllabic information phonetically does not imply that prior perception strategies do not include ''categorizations.'' Categorical perception of certain acoustic continua is well-documented for infants in the first months of life. The contention here is that such perception may be ''acoustic'' only, not involving special phonetic speech processing.

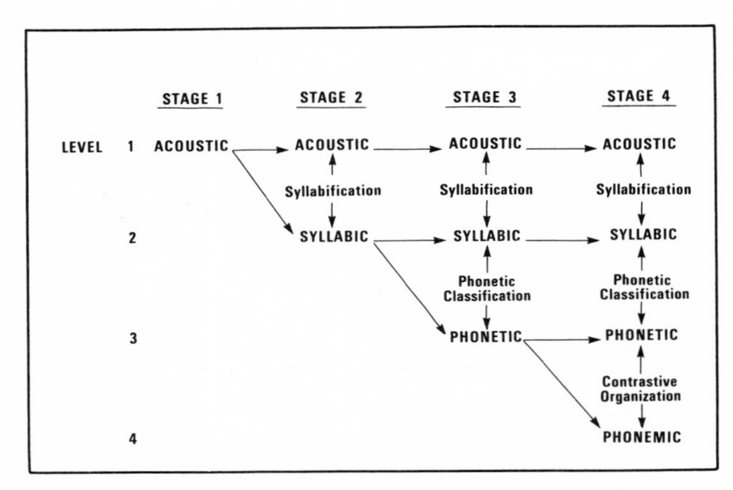

Figure 4. Model of infant speech perception: addition of phonemic level.

This model does *not* suggest that the child has no innate endowment for *developing* a set of linguistic-processing strategies. It does suggest that the achievements of each stage are motivated by real pragmatic needs—the need to work out an efficient strategy for locating units to be analyzed, and the need to keep inventories at manageable sizes. That nature might provide humans with special capacities for arriving at solutions for such needs in speech perception and other cognitive and sensory domains is not surprising. However, the order of occurrence of events posited here is such as to leave open a major role for experience in the child's achievements. For the child to syllabify appropriately, it is necessary for him to know what time frames are appropriate. The time frames could be partially specified innately, but since there is considerable language specificity in syllabic patterns, some tuning is necessary for syllabic timing. The child might determine the appropriate value for syllable timing by observation of a wide variety of rises and falls in power envelopes of speech signals produced by speakers in the child's environment. For the child to phonetically classify appropriately, the child needs to know not only where to look for phonetic units in the acoustic stream (information provided by syllabification), but what certain acoustic patterns have in common with others. Experience with analyzing syllabic patterns over weeks or months may be necessary in order for the child to discover relations among them. For the child to form a system of phonemic contrasts, it is necessary for him to have experience with perceiving and remembering a phonetic inventory that is later summarized and organized as the phonemic system.

This description of what the child presumably must develop in the way of a linguistic-processing system does not, in itself, provide an account of the organismic and experiental factors that influence the system's form and its rate of development. An account of such factors must be included in an explanation of the development of speech perceptual skills. Two basic kinds of factors are involved: (1) those pertaining to the organism and (2) those pertaining to special speech experience.

Organismic factors include (1) anatomical integrity, (2) cognitive capacity (for example, classification skills), and (3) basic, innate auditory-analysis capabilities. These auditory abilities make it possible for the child to perform analyses of amplitude, duration, and spectrum of signals. Whether there are any specific linguistic-processing capacities at the outset remains to be seen. The basic processes of auditory analysis interact with the specific acoustic nature of the code to yield a hierarchy of salience defined jointly by the degree of acoustic difference between the contrastive stimulus pair and by the capacity of the organism to notice that kind of difference.

Experiential factors are related to frequency of occurrence of phonetic and phonemic units and the availability of analogous contrasts for generalization to foreign and infrequently occurring elements.

These factors obviously do not stand alone as predictors of discriminatory behavior. There is both an interaction implied (as in the case of the definition of salience) and a hierarchy of importance in explanation of discrimination data. For instance, the experiential factors may only be important when salience is relatively low. Given high salience and relatively simple stimuli (e.g., monosyllabic forms) analysis may occur primarily at the acoustic level and be little influenced by experiential factors. Presumably, examples of such processing are illustrated by the Eimas *et al.* (1971) work with English voicing contrasts and the Morse (1972) and Till (1976) work with place of articulation. That these discriminations can be shown to be categorical provides a characterization of the basic nature of the auditory-processing system.

Given, however, that auditory salience is relatively low, as in the case of the Spanish-like (or Kikuyu-like) voicing contrasts for stop consonants, experience begins to play a role. Presumably, a high frequency of occurrence of less salient contrasts maximizes the infant's potential for noticing the difference between members of the pair. In the Eilers *et al.* (1979) work, Spanish-learning infants performed significantly better on the Spanish voicing contrast than the English infants. The fact that similar differences favoring English babies were not found for the English-voicing contrast is explained by the relative salience of the English contrast. Experience may have little influence in that case.

A second experiential factor comes into play at least as early as the level of phoneticization. This factor becomes important in explaining how less salient foreign contrasts or less salient, infrequent contrasts become noticed by the infant. If the infant has frequently occurring or more salient, analogous contrasts

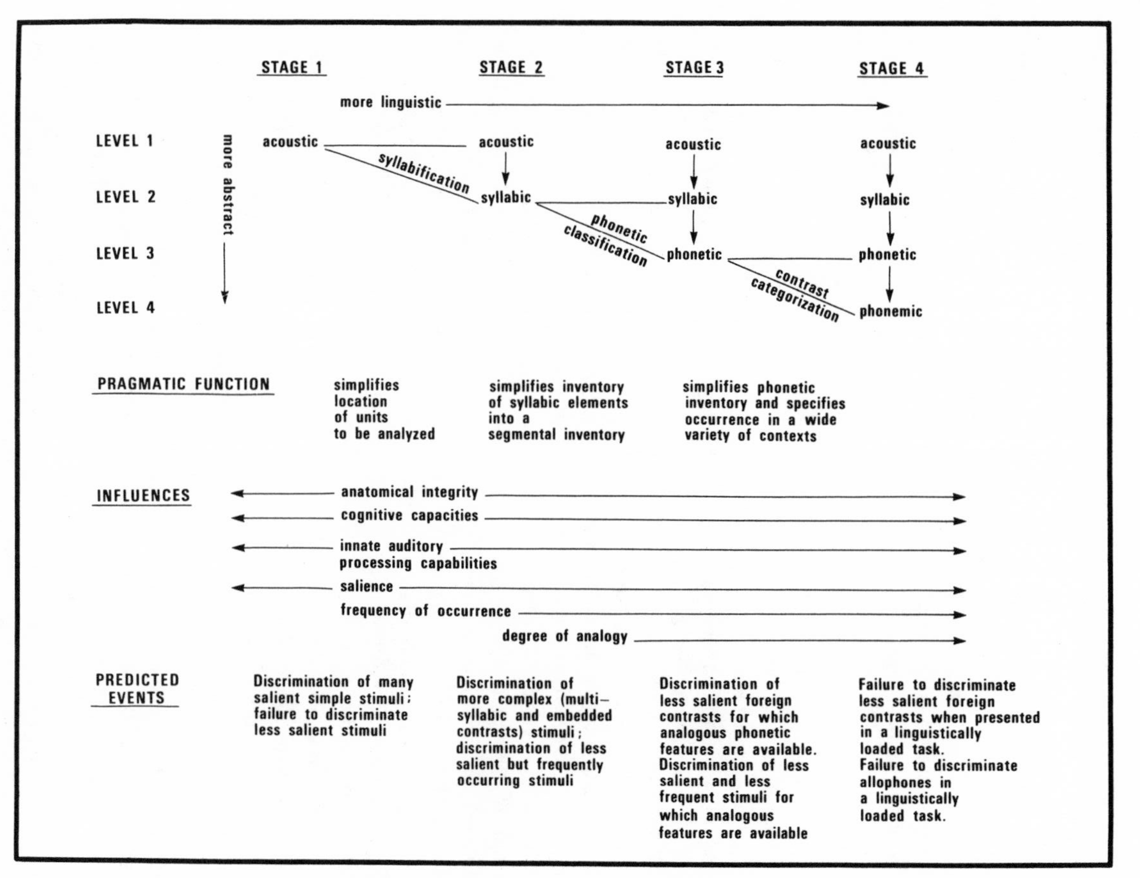

Figure 5. Interaction of organismic and experiential factors with the basic processing model.

to use as models, he may be helped in noticing the specific phonetic features present in the foreign or infrequently occurring native contrast. Thus, he brings a general cognitive process, that of noting similarities among a class of elements (or analogizing), to the speech-perception tasks when a suitable speech substrate exists for him to operate on. The general form of the interaction of organismic and experiential factors with the basic processing model is shown in Figure 5.

Now that we have described the model, it is worthwhile to suggest characteristics of the infant's and young child's behavior that would result from this processing system. Further, it is useful to examine the available literature to see if it meshes well with predictions of the model.

The Data

According to the model, Stage 1 infants have only basic auditory processing skills available to them. These skills are applied to stimulus input and should yield a continuum of success on speech contrasts. Those stimulus inputs that infants succeed in discriminating would be defined as salient and those that infants fail to discriminate would be defined as less salient. The present data base does indicate differing levels of difficulty among different classes of speech sounds. Holmberg, Morgan, and Kuhl (1977) report that babies' ability to discriminate among various fricative consonants differs, a finding previously noted by Eilers and Minifie (1975). Eilers and colleagues derived *Z*-scores indexing discriminability for a variety of speech contrasts and illustrated a continuum of difficulty. Vowel contrasts were more salient (Eilers, Gavin, & Oller, 1980), with certain stop consonants falling lower on the list, even though some were still quite discriminable (Eilers, Morse, Gavin, & Oller, 1981; Morse, Eilers, & Gavin, 1982). At the bottom of the list were certain nasal contrasts (Eilers *et al.*, 1980) and voicing contrasts cued solely by the onset of first-formant voicing (Eilers *et al.*, 1981).

Further features of the model in Stage 1 would include the prediction that infants who are not proficient syllabifiers would have difficulty discriminating multisyllabic stimuli differing in single elements. Jusczyk (1977), reported that infants are able to discriminate [w] and [y] and [d] and [g] in bisyllables in either initial or final position, and Jusczyk and Thompson (1978) found similar results for medial-position stops. Williams and Bush (1978), on the other hand, had to rely on special conditions to show intervocalic discrimination in place of articulation contrasts. Finally, Trehub (1976a,b) demonstrated discrimination of pairs like [apa] versus [aba] but not [atapa] versus [ataba]. The author suggested that the shorter syllables of the 3-syllable stimuli made the discrimination more difficult than in the 2-syllable task where syllable length was greater. Another interpretation of these findings is that acoustic-only processing may not be sufficient for 3-syllable stimuli and that learning appropriate syllabification process may be necessary for processing stimuli with 3 or more syllables, or for bisyllabic stimuli with certain acoustic attributes.

By the time the infant is well into Stage 2, he has probably had a great many hours of listening experience with his native language. His basic auditory apparatus and/or cognitive processing system is beginning to change as a function of noticing high-frequency acoustic events. Part of the reason he is able to notice these events is that he has better strategies for chunking information. He is better at defining syllable-sized units in a wide variety of contexts. At this point, we would expect a reorganization in his salience continuum. Presumably this reorganization moves frequently occurring but less salient contrasts higher up on the continuum so that they become more easily discriminable.

Two sorts of evidence can be brought to bear on this question. The first concerns the Eilers, Wilson, and Moore (1977) finding that discrimination ability improved steadily across the first year of life. Unfortunately, no other studies have looked at infant performance on the same stimuli at two or more ages during the first year of life.

The second kind of evidence comes from cross-linguistic studies. Since frequency of occurrence influences the salience hierarchy, we might expect that infants from different language environments would differ in how the ease of discrimination is ordered. Eimas (1975) and Moffitt (reported in Butterfield & Cairns, 1974) failed to show evidence of discrimination of foreign (i.e., Spanish or Kikuyu-like), minimal (e.g., 30 msec) voicing contrasts in English-learning babies. Streeter (1976) and Lasky *et al.* (1975) demonstrated that Kikuyu and Guatemalan infants, respectively, *could* discriminate minimal voicing contrasts similar to those present in their native language. Later Eilers, Gavin, and Wilson (1979) showed that English-learning infants were poorer discriminators of this foreign contrast than were Spanish babies, in whose native language the contrast had a high frequency of occurrence. Eilers *et al.* (1982) subsequently found that Spanish-learning infants performed significantly better on the Spanish tapped "r" versus trilled "r" contrast than did English-learning babies. Thus, in a variety of experiments, specific language experience affects infant abilities.

According to the model, Stage 3 infants establish a third level of processing. Through the process of phonetic classification, they begin to organize syllabic lists into phonetically oriented syllable types as well as to extract segment and phonetic-feature classifications from the syllable types. Of particular importance at this stage is the infant's ability to use analogy to help organize his phonetic inventory.

Data indicating possible organization by analogy can be found in cross-linguistic studies. The data of particular interest concern speech contrasts that are successfully discriminated by infants from a language background that excludes the contrast. When an infant does discriminate a foreign contrast, we can posit at least two reasons for the success. First, the contrast might be inherently salient, such as many African language contrasts between suction (i.e., click) and non-suction stops. Secondly, the contrast might be similar phonetically to one found in the infant's native language. Even if the foreign contrast is low in salience, the infant might succeed in discriminating it if it is analogous to the native contrast. Eilers *et al.* (1982), found that an [s]–[z] contrast was discriminable by both Spanish- and English-learning infants despite the fact that the contrast does not occur in Spanish. In a previous study, Eilers and Minifie (1975) found an [s]–[z] contrast to be difficult for younger English-learning infants. Thus, [s] versus [z] may be a contrast of low salience. Success by older English-learning infants in discriminating [s] versus [z] may indicate direct learning. The Spanish infants' success, however, would appear to demand a different explanation. When we examine Spanish for a pair analogous to [s] versus [z], we find that Spanish contains a prevoicing contrast [p]–[b] that is highly analogous to that found in English [s] versus [z]. In addition, Spanish has fricative contrasts wherein amplitude (though not voicing) is a fundamental cue. Spanish-learning babies may then use other contrasts as models for discriminating the voiced and voiceless fricatives. Similarly, better performance of Spanish than English babies on a Czech fricative contrast (Eilers *et al.*, 1982) might be accounted for by analogy as well, although the nature of the possible analogy in this case is uncertain.

By Stage 4, the infant or young child is well into what is traditionally thought of as linguistic processing. Through the process of contrast recovery, the child converts phonetic units into phonemic ones tailored to his native language. The child at this stage comes largely to ignore phonetic detail in speech processing. As the process of conversion of phonetic to phonemic units becomes more automatic, the child may have difficulty directly accessing information at the phonetic level.

One manifestation of the processing strategy at this stage would be difficulty in discriminating certain contrasts that were discriminable at earlier stages. At much later stages in development, adults presumably can gain conscious access to information at the phonetic and syllabic levels given appropriate motivation and instruction. This would account for adult abilities to discriminate certain within-phonemic-category contrasts given appropriate tasks and instructions (Carney, Widin, & Viemeister, 1977). Such access may, however, be quite difficult for the child, whose system may not permit such access since it is presumably designed to simplify and streamline an extremely complex task.

Very little work has been done with children between 1 and 2 years of age, and evidence on how they perceive speech is largely unavailable. The fourth level of processing in the present model is influenced by and designed in part to account for the observation that adults and children apparently lose the ability to discriminate some kinds of contrasts that they could presumably discriminate in infancy. For instance, Trehub (1976a) reports that English-speaking adults have great difficulty in discriminating a Czech fricative contrast that English-learning infants *can* discriminate. Miyawaki *et al.* (1975) report that Japanese adults have difficulty discriminating [r] versus [l], an ability that Japanese babies apparently possess. These results suggest that adults come to ignore contrasts that are nonphonemic in their language. Eilers and her associates (Eilers, Gavin, & Wilson, 1979; Eilers, Wilson, & Moore, 1976), on the other hand, demonstrate that English-speaking adults discriminate among VOT contrasts that are within-category or nonphonemic in English. Similar results are reported by Carney *et al.* (1977) and Pisoni, Aslin, Perez, and Hennessy (1978). We might conclude that the discrepancy between adult performance in different studies depends, in part, on the level of processing accessed by subjects, depending on the particular constraints of specific tasks. If instructions and test conditions encourage more linguistic-phonemic processing, adults may be less successful in discriminating nonnative contrasts than if the task encourages access of phonetic, syllabic, or acoustic levels.

To our knowledge, there has been only one study designed to address the possibility that 2- to 3-year-old children discriminate native contrasts better than nonnative ones in real speech. Oller and Eilers (1983) tested English and Spanish 2-year-olds in a procedure called the shell game. Each child was tested on an English [ɹ] versus [w] and a Spanish tap [ɾ] versus trill [r] contrast. The experimenter spoke in the child's native language and taught the child to identify toy objects on instruction. The child was motivated to perform the task by external reinforcers. The child was shown two cups. A candy was placed in one, and both cups were covered with saucers. Two objects were then placed on top of the saucers, one an item familiar to the child (like a rabbit or ladder for the English-learning children and an egg /wéβo/ or car /káro/ for the Spanish-learning children). The children were taught that they could find the candy in one of the cups, if they listened to the experimenter telling them where it was. For example, the experimenter might say, "Look under the one that has the rabbit." Each of the items that was familiar to the child was minimally paired with a nonsense word. The nonsense word was represented to the child as a name for an unfamiliar object. The pairings of words are presented in Table 1.

Each of the word pairs included at least one native word. The paired item was always "nonsense" and half of the time (i.e., in nonnative contrasts) it included a foreign element. Except for the foreign element, the phonotactics were completely Spanish for Spanish-learning children and English for English-learning children. For each contrastive pair, children were given 10 trials with counterbalancing for position and object requested.

Results indicated that the children succeeded on the native contrast and failed on the hybrid or nonnative contrast. We assume that the children failed

Table 1. Shell Game Stimuli for Cross-Linguistic Study

English Contrasts			
Contrastive elements	Real word		Nonsense minimal contrast
ɹ* vs. w*	rabbit [ɹæbɪt]	vs.	[wæbɪt] (native contrast)
ɾ* vs. r	ladder [læɾəɹ]	vs.	[lærəɹ] (hybrid contrast)

Spanish Contrasts			
Contrastive elements	Real word		Nonsense minimal contrast
ɹ vs. w*	huevo [weβo]	vs.	[ɹeβo] (hybrid contrast)
ɾ* vs. r*	carro [karo]	vs.	[kaɾo] (native contrast)

*Denotes native elements.

because they were processing in a linguistic-phonemic mode, a mode in which the hybrid (nonnative) contrasts were ignored. This explanation is particularly appealing since the contrasts were probably both ''salient,'' as suggested by results of previous studies. Both Spanish- and English-learning infants succeeded with an ''r'' versus ''w'' distinction at 6 months of age (Eilers & Oller, 1979) while a tap–trill distinction was shown discriminable by Spanish babies ($p < .03$) and very nearly so by English babies ($p = .056$).

A result in the adult literature by Miyawaki *et al.* (1975) may be similarly interpreted to indicate suppression of nonnative contrasts. Japanese adults were asked to discriminate ''speech syllables'' /ra/–/la/ (an English contrast) and signals that would ''not sound like speech,'' that is, isolated third formants simulating the acoustic difference between the /r/ and /l/ distinction. These Japanese adults performed very accurately on the nonspeech signals, as accurately as English-speaking adults. However, the Japanese adults did not show typical categorical discrimination patterns for the speech signals. These findings suggest that the Japanese adults possess auditory-processing abilities and strategies similar to those of English adults for discriminating ''r'' and ''l'' but that the ability to discriminate a nonnative contrast presented as ''speech'' is suppressed in the phonemic-processing mode.

Conclusion

This chapter attempts to outline certain features of the development of a linguistic-processing system. Rather than asserting that the system is or is not linguistic at any time, our model suggests that the child's system must become *more* linguistic as time passes. Each additional level of processing adds to the linguistic nature of the system. Although there may be specific *linguistic*-perceptual capabilities present at birth, evidence to date is insufficient to demonstrate this possibility.

Current evidence about infant speech perception is supportive of a developmental approach. The multilevel model is offered as a perspective that may provide richer interpretive possibilities than have been available heretofore.

The predictions offered by this model are more elaborate than those presented in our previous work and we look forward to further testing of its hypotheses.

References

Butterfield, E., & Cairns, G. Whether infants perceive linguistically is uncertain, and if they did, its practical importance would be equivocal. In R. L. Schiefelbusch & L. L. Lloyd (Eds.), *Language perspectives: Acquisition, retardation, and intervention.* Baltimore: University Park Press, 1974.

Carney, A. E., Widin, G. P., & Viemeister, N. F. Non-categorical perception of stop consonants differing in VOT. *Journal of the Acoustical Society of America,* 1977, *62,* 961–970.

Eilers, R. E., & Gavin, W. J. The evaluation of infant speech perception skills: Statistical techniques and theory development. In R. Stark (Ed.), *Language behavior in infancy and early childhood.* New York: Elsevier, North Holland, 1981.

Eilers, R. E., & Minifie, F. D. Fricative discrimination in early infancy. *Journal of Speech and Hearing Research,* 1975, *18,* 158–167.

Eilers, R. E., & Oller, D. K. *Cross-linguistic perspective of infant speech perception.* Paper presented at the Society for Research in Child Development, Boston, 1979.

Eilers, R. E., Wilson, W. R., & Moore, J. M. Discrimination of synthetic prevoiced labial stops by infants and adults. *Journal of the Acoustical Society of America,* 1976, *60,* Supplement 1, S91(A).

Eilers, R. E., Wilson, W. R., & Moore, J. M. Developmental changes in speech discrimination by infants. *Journal of Speech and Hearing Research,* 1977, *20,* 766–780.

Eilers, R. E., Gavin, W. J., & Wilson, W. R. Linguistic experience and phonemic perception in infancy: A cross-linguistic study. *Child Development,* 1979, *50,* 14–18.

Eilers, R. E., Wilson, W. R., & Moore, J. M. Speech perception in the language innocent and the language wise: The perception of VOT. *Journal of Child Language,* 1979, *6,* 1–18.

Eilers, R. E., Gavin, W. J., & Oller, D. K. *Cross-linguistic studies in infant speech perception.* Paper presented at the Boston Child Language Conference, October, 1980.

Eilers, R. E., Morse, P. A., Gavin, W. J., & Oller, D. K. The perception of voice-onset-time in infancy. *Journal of the Acoustical Society of America,* 1981, *70,* 955–965.

Eilers, R. E., Gavin, W. J., & Oller, D. K. Cross linguistic perception in infancy: the role of linguistic experience. *Journal of Child Language,* 1982, *9,* 289–302.

Eimas, P. D. Speech perception in early infancy. In L. B. Cohen & P. Salapatek (Eds.), *Infant perception: From sensation to cognition* (Vol. II). New York: Academic Press, 1975.

Eimas, P. D., Siqueland, E., Jusczyk, P., & Vigorito, J. Speech perception in infants. *Science,* 1971, *171,* 303–306.

Holmberg, T. L., Morgan, K. A., & Kuhl, P. A. Speech perception in early infancy: Discrimination of fricative consonants. *Journal of the Acoustical Society of America,* 1977, *62,* Supplement No. 1., S99(A).

Jusczyk, P. W. Perception of syllable-final stop consonants by two-month-old infants. *Perception and Psycholphysics,* 1977, *21,* 450–454.

Jusczyk, P., & Thompson, E. Perception of a phonetic contrast in multisyllable utterances by 2-month-old infants. *Perception and Psychophysics,* 1978, *23,* 105–109.

Kuhl, P. Speech perception in early infancy: Perceptual constancy for vowel categories. *Journal of the Acoustical Society of America,* 1976, *60,* Supplement 1, S90.

Lasky, R. E., Syrdal-Lasky, A. & Klein, R. E. VOT discrimination by four-to-six-and-a-half-month-old infants from Spanish environments. *Journal of Experimental Child Psychology,* 1975, *20,* 215–225.

Miyawaki, K., Strange, W., Verbrugge, R., Liberman, A., Jenkins, J. & Fujimura, O. An effect of linguistic experience: The discrimination of /r/ and /l/ by native speakers of Japanese and English. *Perception and Psychophysics,* 1975, *18,* 331–340.

Moffitt, A. R. Consonant cue perception by twenty- to twenty-four-week-old infants. *Child Development,* 1971, *42,* 717–731.

Morse, P. A. The discrimination of speech and nonspeech stimuli in early infancy. *Journal of Experimental Child Psychology,* 1972, *14,* 477–492.

Morse, P. A., Eilers, R. E. & Gavin, W. J. The discrimination of speech and nonspeech stimuli in early infancy. *Journal of Experimental Child Psychology,* 1972, *14,* 477–492.

Morse, P., Eilers, R. E. & Gavin, W. J. The perception of the sound of silence in infancy. *Child Development,* 1982, *53,* 188–195.

Oller, D. K. & Eilers, R. E. Speech identification in Spanish and English learning 2-year-olds. *Journal of Speech and Hearing Research,* 1983, *26,* 50–53.

Pisoni, D. B., Aslin, R. N., Perey, A. J. & Hennessy, B. L. Identification and discrimination of a new linguistic contrast: Some effects of laboratory training on speech perception. *Research on Speech Perception, Progress Reports* No. 4, Indiana University, 1980.

Streeter, L. A. Language perception of two-month-old infants shows effects of both innate mechanisms and experience. *Nature,* 1976, *259,* 39–41.

Tallal, P. & Stark, R. Perception constancy for phonemic categories: a developmental study with normal and language impaired children. *Journal of Applied Psycholinguistics,* 1980, *1,* 49–64.

Till, J. *Infants' discrimination of speech and non-speech stimuli.* Unpublished doctoral dissertation, University of Iowa, Iowa City, 1976.

Trehub, S. E. The discrimination of foreign speech contrasts by infants and adults. *Child Development,* 1976, *47,* 466–472. (a)

Trehub, S. E. & Jakubovicz, N. *Infants' discrimination of two-syllable stimuli: The role of temporal factors.* Paper presented at meeting of the American Speech and Hearing Association, Houston, Texas, November, 1976.

Trehub, S. E. & Rabinovitch, M. S. Auditory-linguistic sensitivity in early Infancy. *Developmental Psychology,* 1972, *6,* 74–77.

Warren, R. M. & Obusek, C. J. Speech perception and phonemic restoration. *Perception and Psychophysics,* 1971, *9,* 358–362.

Williams, L. & Bush, M. The discrimination by young infants of voiced stop consonants with and without release bursts. *Journal of the Acoustical Society of America,* 1978, *63,* 1223–1225.

CHAPTER 12

Infant Speech Perception

Nature's Contributions

Philip A. Morse

Department of Neuropsychology
New England Rehabilitation Hospital
Woburn, Massachusetts

Nature's contributions to the infant's speech perception abilities were, at one time, a very controversial topic. Some investigators, including the present author, interpreted the *early* findings in infant speech perception as evidence of innate, species-specific linguistic processing, wheras others strongly disagreed with this interpretation (cf. Butterfield & Cairns, 1974; Eimas, 1974b; Morse, 1974). Gradually, over the course of the past decade, evidence has emerged that has helped to clarify the place of nature's contributions in infant speech perception. These studies have revealed an impressive array of speech perception abilities not only in the human infant but also in nonhuman species. Since much of this work has been motivated by research and theory in human *adult* speech perception (particularly by Liberman and his colleagues at Haskins Laboratories), the present chapter begins with the basic findings in adult speech perception that first captured the attention of infant researchers and that subsequently prompted studies of speech perception in nonhuman species. The second section of the chapter discusses the contributions of nature implicated by studies of cross-language infant speech perception. In the third portion of the chapter, nature's contributions are considered within an auditory information-processing framework. In the final section, we return to a familiar issue underlying much of the research in infant speech perception: Which, if any, aspects of speech perception are unique to human "nature?"

The preparation of this chapter was supported by NICHD grants HD-08240 and HD-03352.

Nature's Contributions: From Human Adults to Infants to Nonhumans

Much of the theoretical excitement about the early findings in infant speech perception (e.g., Eimas, Siqueland, Jusczyk, & Vigorito, 1971) was stimulated by the relevance of these data to work in adult speech perception. Studies with adult listeners had revealed that humans (1) perceive stop consonants categorically, (2) exhibit perceptual constancy for phonetic categories (across a variety of acoustic contexts), and (3) demonstrate hemisphere (ear) laterality differences in perceiving some speech sounds. In contrast, nonspeech sounds had been found not to evidence this same pattern of perception. Liberman (Liberman, Cooper, Shankweiler, & Studdert-Kennedy, 1967; Liberman, 1970) suggested that this special mode of processing speech involved an underlying coding of the acoustic information that was based on the *production* of speech sounds (a motor theory of speech perception). Evidence in human infants for categorical perception, perceptual constancy, and hemispheric specialization would suggest that processing in the speech mode is not learned through extensive language experience, but that humans are probably born with these abilities. Furthermore, since only humans can produce the sounds of speech, a motor theory of speech perception would predict that nonhumans would fail to exhibit the speech perception abilities found in human adults and infants. Armed with these theoretical predictions, and the empirical findings and stimuli from adult research, investigators began to study these abilities first in infants and then in nonhumans.

Categorical Perception

Studies of categorical perception have demonstrated that adult listeners discriminate differences in stimuli along an acoustic continuum to the extent that they can assign different phonetic labels to the stimuli. Thus, when presented with stimuli varying along a voicing continuum (e.g., [ba] versus [pa] cued by differences in voice-onset-time) or along a place of articulation continuum (e.g., [ba] versus [da] versus [ga] cued by differences in starting frequency of the second formant transition), adult listeners discriminate pairs of stimuli that they label *differently* (between-category contrasts), but do relatively poorly in discriminating a comparable acoustic difference along the continuum in stimuli selected from the *same* phonetic category (within-category contrasts).

Categorical discrimination has been studied in infants by presenting subjects with between-category versus within-category contrasts selected according to the adult category labeling data. In the first of these experiments, Eimas *et al.* (1971) demonstrated categorical discrimination in one- and four-month-old infants using the voicing contrast [ba] versus [pa]. Miller (1974) and Eilers, Wilson, and Moore (1979) subsequently replicated this finding, and Eimas (1975b) extended it to the voicing contrast [da] versus [ta]. Categorical discrimination has also been observed in infants for stimuli along place of articulation ([b], [d], [g]) continua (Eimas, 1974a; Miller & Morse, 1976; Till, 1976), for the liquid contrasts [ra] versus [la] (Eimas, 1975a), for the stop–nasal continuum of [ba] versus [ma] (Eimas & Miller, 1980b), and for a stop–semivowal continuum (Eimas & Miller, 1980a). These studies clearly demonstrate that infants exhibit a pattern of categorical discrimination for a wide range of consonant contrasts that is similar to that observed in adult listeners. To date, the only major discrepancy between the infant and adult data appears to be the lack of evidence of categorical discrimination for fricatives in infants (Jusczyk, Murray, & Bayley, 1979). Infants' perception of vowels, which are perceived more continuously rather than categorically, is also similar to that found in adult listeners. Relatively long duration tokens of the vowels [i] and [I] are discriminated continuously (Swoboda, Morse, & Leavitt, 1976), whereas more categorical-like discrimination characterizes the infant's perception of briefer versions of the same vowels (Swoboda, Kass, Morse, & Leavitt, 1978). In sum, these infant studies of categorical and continuous discrimination demonstrate that most of these speech perception abilities are evident in humans within the first few postnatal months, prior to the age at which infants experience these contrasts in their own speech production.

Perceptual Constancy

The second body of adult data that suggests that speech sounds are processed in a special mode is the research on "perceptual constancy." A well-known example of this phenomenon is the syllable contrast [di] versus [du] in which the critical cue for [d] is conditioned by the vowel context in which the consonant occurs (Liberman, 1970). The syllable [di] contains a rising second-formant transition, whereas a falling second-formant transition is the critical cue for [du]. Yet, despite this variation in the contextual information, listeners identify the same initial consonant in both syllables. Thus, in this example, the listener is able to decode or abstract a constant percept [d] from the variation in acoustic information. Studies of perceptual constancy in infants have differed considerably in the extent to which this variation in acoustic information is merely irrelevant and must be ignored by the listener as opposed to decoded in order to extract the underlying constant percept. Further, since the stimuli employed in these studies have often been natural speech sounds, the acoustic cues have been less clearly specified or controlled than in studies of categorical discrimination. Nevertheless, these studies suggest that at least by 6 months of age infants do evidence some forms of perceptual constancy. The first such study was conducted by Fodor, Garrett, and Brill (1975) using the stımuli [pi], [ka], and [pu]. Infants' learning of a conditioned head-turning response was found to be superior if [pi] and [pu] were paired against [ka], than if [pi] and [ka] were paired against [pu]. This result has been interpreted as evidence of perceptual constancy for the category [p] in the context of varying vowel environments. In our own laboratory, we attempted to demonstrate perceptual constancy for [di] versus [du] using a conditioned head-turning procedure (Morse, Leavitt, Donovan, Kolton, Miller, & Judd-Engel, 1975) and for the consonant categories [b] versus [g] with five different vowel environments using the nonnutritive sucking paradigm (Morse & Suomi, 1979). Unfortunately, the multiple-training session procedure of the first study required too much of the infant, whereas the vowel variation in the second study proved too interesting for the infants and they failed to habituate during the first phase of the test procedure. To date no studies of perceptual constancy have been carried out with voicing contrasts, but Kuhl and her colleagues have investigated perceptual constancy for the fricatives [f] versus [θ] and [s] versus [ʃ] in initial and final positions (Holmberg, Morgan, & Kuhl, 1977; Kuhl, 1980), for the manner contrast [m, n, ŋ] versus [b, d, g] (Hillenbrand, 1980), and for the vowels [a] versus [ɔ] (Kuhl, 1977) and [i] versus [a] (Kuhl, 1979b; Kuhl & Miller, 1982). In these studies, infants approximately 6 months of age were tested using a head-turning paradigm in which they were trained over several sessions to discriminate phonetic categories that contained successive irrelevant acoustic information, for example, variations in speaker, intonation contour, or vowel context (in the case of consonants). Kuhl and her colleagues observed that infants learned to make all of these perceptual constancy discriminations. Furthermore, control measures indicated that these findings were not due to the infants merely memorizing sets of speech sounds that were reinforced by the experimenter. In summary, the infant research on perceptual constancy suggests that infants do demonstrate some of the same patterns of perceptual constancy for speech sounds observed in adult listeners. However, additional research is needed, especially with synthetic stimuli, to determine the acoustic bases of the types of constancy exhibited by young infants.

Hemispheric Laterality

Studies of adult speech perception have also shown that many speech sounds (especially stop consonants) are processed differentially in the left hemisphere. In dichotic listening studies, when adult listeners are presented simultaneously with different speech sounds in each ear, they tend to exhibit a slight but reliable right-ear (left hemisphere) advantage in identifying the two sounds. Studies using auditory averaged evoked potentials have demonstrated differential electrophysiological responses to speech over the left hemisphere and to musical tones over the right hemisphere.

Infant studies have also employed these behavioral and evoked-potential techniques in examining the processing of speech and nonspeech sounds. In the first behavioral study of hemispheric differences using a nonnutritive sucking procedure, Entus (1977) found that infants 3 to 20 weeks of age could discriminate a dichotically presented speech contrast better in the right ear and a nonspeech (musical note) contrast better in the left ear (right hemisphere). Although a subsequent study by Vargha-Khadem and Corballis (1977) failed to replicate this finding, Best and her co-workers, using a heart-rate procedure did observe a right-ear advantage for place discrimination in 3- and 4-month-olds and a left-ear advantage for music in infants 2, 3, and 4 months of age (Best, Hoffman, & Glanville, 1979; Glanville, Best, & Levenson, 1977).

Auditory evoked-potential (AEP) studies in infants have also provided evidence of hemispheric differences in speech processing at an early age. Molfese observed that newborns, 5-month-olds, 1-year-olds, and adults responded with greater AEP amplitudes over the left hemisphere (LH) to words and speech sounds and with greater AEP amplitudes over the right hemisphere (RH) to nonspeech sounds (Molfese, Freeman, & Palermo, 1975; Molfese, 1977). More recently, Molfese and his co-workers have examined components of the AEP over the LH and RH using place of articulation and voicing contrasts. These studies revealed that for place of articulation contrasts a component of the AEP over the *LH* varied with changes in formant transitions for adults, newborns, and even preterm infants (Molfese, 1978; Molfese & Molfese, 1979a; Molfese & Molfese, 1980). In addition, a component of the LH AEP of adults and infants has even been observed to reflect some degree of perceptual constancy for the consonants [b] and [g] independent of vowel context (Molfese, 1978, 1980). On the other hand, studies of voicing contrasts have consistently yielded a reliable *RH* component that varies with voice-onset-time category differences in adults, 4-year-old children, 2-month-olds, but not in newborns (Molfese, 1978; Molfese & Hess, 1978; Molfese & Molfese, 1979b). Although the AEP findings of RH effects for voicing and LH effects for place of articulation appear to be inconsistent with the LH dichotic listening effects for both types of contrasts in adults, Molfese and his colleagues have observed this pattern of laterality differences consistently across a wide developmental range. Finally, a RH component has also been observed for nonspeech differences in relative onset of two tones (tone-onset-time) in adults and children (Molfese, Erwin, & Deen, 1980; Molfese, 1983). In summary, studies of hemispheric laterality effects indicate that the behavioral and electrophysiological patterns of hemispheric differences observed in adult listeners are also found in young infants, and in some instances even in newborns and preterm infants.

Taken together, these studies of infant speech perception consistently demonstrate a general pattern of categorical discrimination, perceptual constancy, and hemispheric differences that is similar to that observed in adult listeners. Neurophysiological research with nonhuman species has revealed that individual neurons or feature detectors in the primary sensory cortices of the monkey and cat respond selectively to auditory and visual stimulus features that have special significance for these species. Given the species-specific nature of speech for humans, human adults and infants may possess feature detectors in their auditory systems that respond selectively to the acoustic and phonetic features of speech sounds and underlie some of these perceptual abilities (Eimas & Tartter, 1979). Eimas and others (e.g., Eimas & Corbit, 1973; Eimas, Cooper, & Corbit, 1973) have, in fact, gathered behavioral evidence in support of the adaptation of feature detectors for speech in humans. However, more recent speech adaptation studies have questioned whether acoustic rather than phonetic feature detectors are being adapted and whether adaptation involves changes in response biases rather than perceptual shifts (e.g., Ades, 1976). Nevertheless, these feature-adaptation results are generally consistent with the hypothesis that the human auditory system is structured to respond to some features of human vocalizations. Although this structuring of the human auditory system may underlie the speech perception abilities observed in human adults and infants, only studies with *nonhuman* listeners can determine if these feature-adaptation effects and the patterns of categorical discrimination, perceptual constancy, and hemispheric differences observed in infants are unique to human "nature."

Although speech perception research with nonhuman listeners has not included work on feature adaptation, it has explored categorical discrimination, perceptual constancy, and hemisphere laterality effects. For example, Morse and Snowdon (1975) demonstrated that rhesus monkeys exhibit between-category discrimination for place of articulation contrasts that is superior to within-category performance. Similar nonhuman primate results have been reported for place of articulation contrasts by Sinnott, Beecher, Moody, and Stebbins (1975) and Kuhl and Padden (1983a) and for voicing contrasts (Kuhl & Padden, 1983b). However, research with chinchillas has revealed that categorical discrimination may be observed in nonprimate mammals as well (Kuhl, 1976, 1979; Kuhl & Miller, 1975, 1978). Hemisphere laterality studies of categorical discrimination have demonstrated that nonhuman and adult human primates exhibit similar patterns of electrophysiological asymmetry in discriminating voicing contrasts (Morse, Molfese, Laughlin, Linnville, & Wetzel, submitted). However, in one study nonhuman and adult human primates were found to differ in their patterns of lateralized categorical discrimination of place contrasts (Molfese, Laughlin, Morse, Linnville, & Wetzel, submitted).

In general, most of the nonhuman speech perception results to date suggest that the pattern observed in the human infant's speech perception abilities is *not* unique to our species and therefore may not be attributable to our special abilities in producing speech. Instead, these human infant speech perception abilities appear to reflect the general acoustic processing abilities of the primate or mammalian auditory system. Although enough studies of cross-species perception have not yet been carried out to offer strong support for this position, the monkey and chinchilla research on basic categorical discrimination clearly reveals that this feature of processing speech is not special to adult or infant human nature, but instead reflects ''natural'' constraints of the primate, and more generally, mammalian auditory system.

In sum, the human infant and animal speech perception data have revealed that the major aspects of processing speech observed in adult humans may indeed be thought of as ''innate'' and observable in humans with very little experience with speech. However, their innateness is not restricted to *Homo sapiens,* for all of these aspects of speech processing are in some instance observable in nonhuman primates and, in the case of categorical discrimination, in nonprimate mammals as well. Since these animal data reveal that the ability to produce human speech sounds does not provide the basis for *these* features of speech perception, one might wonder whether there are any *other* aspects of speech perception that are based on the production of speech, and, if so, whether they are unique to our species. It has been suggested that adults' perception of silence between phonetic segments, such as in the contrast [slIt] versus [splIt], may reflect an appreciation of the *production* of strings of speech sounds (Dorman, Raphael, & Liberman, 1979). A recent study carried out in collaboration with Rebecca Eilers (Morse, Eilers, & Gavin, 1982) indicates that infants 6 months of age discriminate silence in the contrasts [slIt] versus [splIt] in a manner similar to that observed in adult listeners. However, since this study was conducted with natural speech sounds, it did not permit a more detailed assessment of whether infants treat spectral and silence cues in this contrast in a manner similar to adults. This work is currently in progress by Best and her colleagues (personal communication). In the interim, the Morse *et al.* (1982) study suggests that if this knowledge of the constraints on speech production plays a role in speech perception, it does so without the benefit of much, if any, vocal or acoustic developmental experience. Will the appropriate animal experiments also reveal that this variant of a motor theory of speech perception does not account for these perceptual findings and that instead these speech perception abilities are also based on properties of the primate or mammalian auditory system?

Cross-Language Studies: The Importance of Salience

If the infant's impressive speech perception abilities reviewed above are not based on ontogenetic or phylogenetic experience in producing the sounds of speech, what role, if any, does *perceptual* experience play? Investigators have explored this question by presenting infants with language contrasts that do not occur in the language to which they are primarily exposed. Some of these studies clearly demonstrate the importance of perceptual experience in discriminating certain speech contrasts (cf. Chapter 11). However, for other speech contrasts, discrimination appears not to depend on perceptual experience.

Studies of voicing discrimination were the first to suggest that infants could discriminate phonetic contrasts that do not occur in their native languages. The adult research on the voiced–voiceless distinctions of different languages has revealed that there are primarily three categories that most languages employ along the voice-onset-time continuum. We shall refer to these three categories as the ''prevoiced,'' ''voiced,'' and ''voiceless'' categories. English employs the voiced and voiceless categories for its voicing contrasts, for example [ba] versus [pa], whereas Spanish, Kikuyu (a Bantu language), and many other languages use the prevoiced and voiced categories to signal voicing contrasts. When Spanish-learning and Kikuyu-learning infants are presented with the English-voicing contrast (which does not occur in their native languages), they nevertheless *are* able to discriminate across the English [ba] versus [pa] category boundary (Eilers, Gavin, & Wilson, 1979; Lasky, Syrdal-Lasky, & Klein, 1975; Streeter, 1976). On the other hand, Spanish-learning infants, for whom the prevoiced/voiced contrast is native, discriminate small differences across that voicing boundary, whereas English-learning infants do not (Eilers, Gavin, & Wilson, 1979). It appears from these studies that discrimination across the English-voicing boundary does not require language-specific experience, whereas the opposite is true for the Spanish-voicing boundary. However, closer examination of the synthetic stimuli employed in these two sets of contrasts suggests that the Spanish-voicing contrast is cued by a single acoustic difference (relative onset of voicing, i.e., onset of the first formant relative to the onset of the other formants), whereas the English contrast is signaled by other cues as well (duration and extent of the first formant transition, amount of aspiration). Thus, the reason that discrimination across the English-voicing contrast may be independent of language-specific experience is that it may be more salient for infants due to the availability of these multiple cues (Stevens & Klatt, 1974).

In order to explore this possibility, Eilers, Morse, Gavin, and Oller (1981) presented 6-month-old infants with natural and synthetic [ba] versus [pa] or [du] versus [tu] voicing contrasts. One group of infants received the [ba] versus [pa] pairs, and a second group was presented with the [du] versus [tu] contrast. Both synthetic contrasts differed by 20 msec across the adult category boundary and were discriminable by adult subjects. The synthetic [ba] versus [pa] contrast contained the multiple acoustic cues mentioned above, whereas synthetic [du] versus [tu] differed solely in the onset of voicing. Infants were able to discriminate both natural contrasts (both of which contained multiple and large acoustic differences), but only the multiple-cued [ba] versus [pa] synthetic contrast was discriminated by the infant listeners. These results suggest that the specific acoustic cues employed do affect the infant's discriminability of voicing contrasts at the English boundary. They further suggest that studies of infant cross-language perception need to consider the relative salience of the cues employed, before the role of perceptual experience can be assessed. Two other studies in the literature may be interpreted in a similar manner. Trehub (1976) reported that English-learning infants could discriminate the Czech contrast [za] versus [řa] and Werker, Gilbert, Humphrey, and Tees (1981) found that 7-month-old English-learning infants were capable of discriminating two Hindi contrasts (dental vs. retroflex and aspirated vs. unaspirated voiced stops). In both studies, adult native speakers were also able to discriminate these phonetic differences, but English-speaking adults with no previous exposure to these contrasts failed to discriminate them. Thus, *these* cross-language infant studies indicate that some phonetic contrasts contain acoustic cues that are sufficiently salient to permit the auditory systems of young infants to discriminate them without previous experience with these contrasts. However, adult listeners, who lack experience with these contrasts and have experience with other contrasts instead, appear to have considerable difficulty discriminating these same speech sounds. Therefore, what may be salient for the inexperienced auditory system of the young infant (or nonhuman primate), may no longer be for phonological and/or neurophysiological reasons for the sophisticated and over-experienced adult.

Auditory Information Processing

Nature's contributions to infant speech perception are not restricted to the properties of the auditory system as they affect categorical discrimination, perceptual constancy, hemispheric asymmetries, or the discrimination of perceptually salient speech contrasts. Nature also constrains the general information processing of auditory and speech events by the infant. In models of adult information processing, auditory information is thought to be initially retained in a very brief preperceptual memory or echoic store, during which the critical features of the stimulus are recognized by the listener. Subsequent information-processing stages include a memory store of limited size and duration (short-term memory) and a more extensive, unlimited long-term memory store. If the infant is to be able to recognize, discriminate, and remember speech sounds in the typical discrimination paradigms employed or in the running speech of adults, he or she presumably makes use of these or similar stages of auditory information processing. Although investigators have developed discrimination paradigms that successfully operate within the infant's information-processing abilities, very little research directly explores these abilities and their development.

Research on the earliest stage of auditory information processing, echoic memory, has only recently begun. Cowan, Suomi, and Morse (1982) conducted a series of studies with 8-week-old infants using a modified nonnutritive sucking procedure and forward- and backward-masking paradigms. Research with adult listeners had demonstrated that at brief interstimulus intervals (e.g., 50-msec stimulus-onset-asynchrony or SOA), the second sound in a pair interferes with the processing of the first sound (backward masking) more than the first sound will interfere with the recognition of the second one (forward masking). However, after 250-msec SOA, adult listeners exhibit asymptotic performance in backward-masking tasks, gaining little from longer processing time before the onset of the second sound. Cowan *et al.* presented infants in one study with two vowels separated by 50-msec SOA in a forward-masking, backward-masking, or control condition. As expected from the adult literature, infants did much better at discriminating a vowel change in the forward-masking condition (performance in the backward-masking condition did not differ from a no-change control group). In a second study with 250-msec and 400-msec SOA backward-masking conditions, infants required 400-msec of SOA before they could discriminate this same vowel contrast, whereas adults needed only 250 msec, and their performance did not improve with additional processing time. These results suggest that the useful duration of echoic memory for infants may be somewhat longer than that found in adult listeners. A similar finding has recently been reported for the earliest stage of visual information processing (Lasky & Spiro, 1980).

In contrast to the extensive research on infant *visual* short-term memory (STM), studies of the infant's auditory STM abilities for speech are remarkably sparse. Two series of investigations in our own laboratory have provided some data on this stage of information processing. In one series of vowel studies, we found that infants discriminated 240-msec tokens of the vowels [i] versus [I] in a continuous manner (good between- *and* within-category discrimination), but when these vowels were only 60 msec in duration their within-category performance became less reliable, yielding a somewhat more categorical-like pattern of discrimination (Swoboda *et al.*, 1976, 1978). This finding is similar to the results observed in adult listeners, which have been interpreted as evidence for the contributions of phonetic STM and auditory STM in the perception of vowels. (Since auditory STM is less available to the listener in brief duration vowels, this results in poorer within-category discrimination.) In another series of studies in our laboratory, we presented young infants with two different cardiac-orienting–response-discrimination paradigms: one in which the stimulus was blocked in trials with long (approximately 30 sec) intertrial intervals and one in which no intertrial interval was employed and stimuli were strung together in long sequences. In two separate studies (Leavitt, Brown, Morse, & Graham, 1976;

Miller, Morse, & Dorman, 1977), it was observed that young infants were able to discriminate a speech contrast in the paradigm without the long intervals, but *not* in the one that contained long intervals between trials, which is successful with older infants (Moffitt, 1971).

Studies of long-term auditory memory in infancy are virtually nonexistent. This is particularly surprising in view of the fact that the infant must rely on long-term memory for speech sounds and, in particular, specific combinations of speech sounds to begin to recognize some meaning and order in the speech of caretakers. Investigators who test 6-month-old infants at weekly intervals with a conditioned head-turning paradigm often observe that infants remember specific syllables from week to week, but little systematic work has been done on the infant's long-term memory abilities for speech or other auditory stimuli.

The information-processing tasks that confront the infant, who is listening to speech and attempting to make some sense out of it, extend beyond these stages of processing. The infant is also faced with the segmentation of strings of speech sounds and the recognition of familiar patterns (words) within these strings. In a recent study conducted in our laboratory, Goodsitt, Morse, Cowan, and VerHoeve (1984) explored the 6-month-old infant's ability to recognize a familiar contrast (learned in a previous training session) when it was embedded in a string of two other syllables. We wondered if the *position* in the string of this target contrast would affect recognition and/or if infants would perform better if the other two syllables in which the targets were embedded were identical to each other (redundant condition) or different (mixed condition). The results of two experiments revealed no differences in recognition of a familiar syllable presented at the beginning, middle, or end of the string, but recognition was facilitated by *less* variability in the string in which the targets were embedded (redundant condition). This finding suggests that the structure or pattern of speech sounds that constitute the sequence of utterances spoken to young infants is an important factor and processing constraint in the infant's ability to utilize his or her impressive speech discriminatory capacities.

Infant Speech Perception and Human Nature

Historically, research in infant speech perception began by inquiring if infants exhibit the same perceptual phenomena as adult listeners. Although the findings revealed that very young infants were indeed capable of categorical discrimination, perceptual constancy, and hemispheric asymmetries in their perception of speech, studies with nonhuman listeners suggested that these abilities might be due to properties of the primate or mammalian auditory systems and not limited to ''human nature.'' Furthermore, the cross-language findings reviewed above suggest that the auditory system of the young infant (and presumably of the nonhuman listener as well) finds some acoustic cues more salient and discriminable in phonetic contrasts than other cues. The limited research on auditory information processing leads us to suspect that infants probably process speech and other auditory information in a qualitatively similar manner to adults, although there may be quantitative differences. Appropriate animal comparison studies have not been conducted, but it would be of no great surprise to discover that the auditory systems of primates and other mammals process auditory information in a manner similar to the stages of processing evidenced in human infants and adults. Thus, if all or most of these speech processing abilities prove to be characteristic of nonhuman primates or mammals, which, if any, aspects of speech perception are unique to humans and thus part of only *our* nature?

One area of research that merits consideration is the relative ease with which humans form associations between auditory and visual inputs. This is something humans do quite easily when they learn to associate arbitrary strings of speech sounds that we refer to as ''words'' or ''names'' with their visual referents. The research available on the abilities of animals to understand auditorily the names for objects is limited, but what little work that has been conducted with dogs (e.g., Gilman, 1921; Warden & Warner, 1928) suggests that even highly trained dogs have difficulty responding correctly to an auditory name for an object in a multiple-choice task, although they are able to carry out an impressive series of full-body commands (e.g., ''roll over''). Recently, Campos and his colleagues (Thomas, Campos, Shucard, Ramsay, and Shucard, 1981) investigated the development of this type of cross-model matching in infants. They studied the 11- and 13-month-old's comprehension of concrete nouns by observing direction of eye fixations to the referent of an object word spoken by the mother. Infants 11 months of age failed to look reliably at the appropriate object in response to a word that the mother felt her infant knew. However, the 13-month-old infants were able to direct their looking significantly longer at the referent of the known word. These results suggest that, at approximately one year of age, the infant begins to associate sequences of speech sounds spoken by caregivers with objects in his environment. This ability may depend on developmental changes in auditory and visual association cortex and/or general cognitive or communicative changes at this age. Additional research is necessary to determine how refined these discriminative and cross-model matching abilities are at this age and how they develop into adultlike ''names.'' Furthermore, since dogs appear to have considerable difficulty acquiring these auditory-visual associations, it would be of great interest to extend these investigations to nonhuman primates. Until these comparative experiments are carried out, we may safely fantasize that one aspect of the infant's speech perception abilities that may be unique to human nature is the use of these impressive discriminative and classificatory abilities in the service of understanding words and names for objects in his environment.

References

Ades, A. E. Adapting the feature detectors for speech perception. In E. C. T. Walker & R. J. Wales (Eds.), *New approaches to language mechanisms*. The Hague: North Holland, 1976.

Best, C., Hoffman, H., & Glanville, B. Brain lateralization in 2-, 3-, and 4-month-olds for phonetic and musical timbre discriminations under memory load. *Haskins Laboratories Status Report on Speech Research*, 1979, *SR-59/60*, 1–30.

Butterfield, E., & Cairns, G. Discussion summary—Infant reception research. In R. Schiefelbusch & L. Lloyd (Eds.), *Language perspectives—acquisition, retardation, and intervention*. Baltimore: University Park Press, 1974.

Cowan, N., Suomi, K., & Morse, P. Echoic storage in infant perception. *Child Development*, 1982, *53*, 984–990.

Dorman, M., Raphael, L., & Liberman, A. Some experiments on the sound of silence in phonetic perception. *Journal of the Acoustical Society of America*, 1979, *65*, 1518–1532.

Eilers, R., Gavin, W., & Wilson, W. Linguistic experience and phonemic perception in infancy: A cross-linguistic study. *Child Development*, 1979, *50*, 14–18.

Eilers, R., Morse, P., Gavin, W., & Oller, D. K. Discrimination of voice onset time in infancy. *Journal of the Acoustical Society of America*, 1981, *70*, 955–965.

Eilers, R., Wilson, W., & Moore, J. Speech discrimination in the language-innocent and language-wise: A study in the perception of voice-onset-time. *Journal of Child Language*, 1979, *6*, 1–18.

Eimas, P. Auditory and linguistic processing of cues for place of articulation by infants. *Perception and Psychophysics*, 1974, *16*, 513–521. (a)

Eimas, P. Infant speech perception research. In R. Schiefelbusch & L. Lloyd (Eds.), *Language perspectives—acquisition, retardation, and intervention*. Baltimore: University Park Press, 1974. (b)

Eimas, P. Auditory and phonetic coding of the cues for speech: Discrimination of the [r-l] distinction by young infants. *Perception and Psychophysics*, 1975, *18*, 341–347. (a)

Eimas, P. Speech perception in infancy. In L. Cohen & P. Salapatek (Eds.) *Infant perception: From sensation to cognition* (Vol. II). New York: Academic Press, 1975. (b)

Eimas, P., Cooper, W., & Corbit, J. Some properties of linguistic feature detectors. *Perception and Psychophysics*, 1973, *13*, 247–252.

Eimas, P., & Corbit, J. Selective adaptation of linguistic feature detectors. *Cognitive Psychology*, 1973, *4*, 99–109.

Eimas, P., & Miller, J. Contextual effects in infant speech perception. *Science*, 1980, *209*, 1140–1141.

Eimas, P., & Miller, J. Discrimination of the information for manner of articulation. *Infant Behavior and Development*, 1980, *3*, 367–375. (b)

Eimas, P., Siqueland, E., Jusczyk, P., & Vigorito, J. Speech perception in infants. *Science*, 1971, *171*, 303–306.

Eimas, P. D., & Tartter, V. C. On the development of speech perception: Mechanisms and analogies. In H. W. Reese & L. P. Lipsitt (Eds.), *Advances in child development and behavior* (Vol. 13). New York: Academic Press, 1979.

Entus, A. Hemispheric asymmetry in processing of dichotically presented speech and nonspeech stimuli by infants. In S. Segalowitz & F. Gruber (Eds.), *Language development and neurological theory*. New York: Academic Press, 1977.

Fodor, J., Garrett, M., & Brill, S. Pi ka pu: The perception of speech sounds by prelinguistic infants. *Perception and Psychophysics*, 1975, *18*, 74–78.

Gilman, E. A dog's diary. *Journal of Comparative Psychology*, 1921, *1*, 309–315.

Glanville, B., Best, C., & Levenson, R. A cardiac measure of cerebral asymmetries in infant auditory perception. *Developmental Psychology*, 1977, *13*, 54–59.

Goodsitt, J., Morse, P., Cowan, N., & VerHoeve, J. Infant speech recognition in multisyllabic contexts. *Child Development*, 1984, *55*, 903–910.

Hillenbrand, J. *Perceptual organization of speech sounds by young infants*. Unpublished Ph.D. dissertation, University of Washington, Seattle, Washington, 1980.

Holmberg, T., Morgan, K., & Kuhl, P. Speech perception in early infancy: Discrimination of fricative contrasts. *Journal of Acoustical Society of America*, 1977, *62* (Suppl. 1), S99(A).

Jusczyk, P., Murray, J., & Bayley, J. *Perception of place of articulation in fricatives and stops by infants*. Paper presented at the biennial meeting of the Society for Research in Child Development, San Francisco, California, 1979.

Kuhl, P. Speech perception by the chinchilla: Categorical perception of synthetic alveolar plosive consonants. *Journal of the Acoustical Society of America*, 1976, *60* (Suppl. 1), S81(A).

Kuhl, P. Speech perception in early infancy: Perceptual constancy for the vowel categories /a/ and /ɔ/. *Journal of the Acoustical Society of America*, 1977, *61* (Suppl. 1), S39(A).

Kuhl, P. Models and mechanisms in speech perception. *Brain, Behavior, and Evolution*, 1979, *16*, 374–408. (a)

Kuhl, P. Speech perception in early infancy: Perceptual constancy for spectrally dissimilar vowel categories. *Journal of the Acoustical Society of America*, 1979, *66*, 1668–1679. (b)

Kuhl, P. Perceptual constancy for speech sound categories. In G. Yeni-Komshian, J. Kavanagh, & C. Ferguson (Eds.), *Child phonology: Perception and production*. New York: Academic Press, 1980.

Kuhl, P., & Miller, J. Speech perception by the chinchilla: Voiced-voiceless distinction in alveolar plosive consonants. *Science*, 1975, *190*, 69–72.

Kuhl, P., & Miller, J. Speech perception by the chinchilla: Identification functions for synthetic VOT stimuli. *Journal of the Acoustical Society of America*, 1978, *63*, 905–917.

Kuhl, P., & Miller, J. Discrimination of auditory target dimensions in the presence or absence of orthogonal variation in a second dimension by infants. *Perception and Psychophysics*, 1982, *31*, 279–292.

Kuhl, P., & Padden, D. Enhanced discriminability at the phonetic boundaries for the place feature in macaques. *Journal of the Acoustical Society of America*, 1983, *73*, 1003–1008. (a)

Kuhl, P. & Padden, D. Enhanced discriminability at the phonetic boundaries for the voicing feature in macaques. *Perception and Psychophysics*, 1983, *32*, 542–550. (b)

Lasky, R., & Spiro, D. The processing of tachistoscopically presented visual stimuli by five-month-old infants. *Child Development*, 1980, *51*, 1292–1294.

Lasky, R., Syrdal-Lasky, A., & Klein, R. VOT discrimination by four and six and a half month old infants from Spanish environments. *Journal of Experimental Child Psychology*, 1975, *20*, 215–225.

Leavitt, L., Brown, J., Morse, P., & Graham, F. Cardiac orienting and auditory discrimination in 6-week infants. *Developmental Psychology*, 1976, *12*, 514–523.

Liberman, A. The grammars of speech and language. *Cognitive Psychology*, 1970, *1*, 301–323.

Liberman, A., Cooper, F., Shankweiler, D., & Studdert-Kennedy, M. Perception of the speech code. *Psychological Review*, 1967, *74*, 431–461.

Miller, C., & Morse, P. The 'heart' of categorical speech discrimination in young infants. *Journal of Speech and Hearing Research,* 1976, *19,* 578–589.

Miller, C., Morse, P., & Dorman, M. Cardiac indices of infant speech perception: Orienting and burst discrimination. *Quarterly Journal of Experimental Psychology,* 1977, *29,* 533–545.

Miller, J. *Phonetic determination of infant speech perception.* Unpublished Ph.D. dissertation. University of Minnesota, Minneapolis, Minn., 1974.

Moffitt, A. Consonant cue perception by twenty- to twenty-four-week old infants. *Child Development,* 1971, *42,* 717–731.

Molfese, D. L. Infant cerebral asymmetry. In S. Segalowicz & F. Gruber (Eds.), *Language development and neurological theory.* New York: Academic Press, 1977.

Molfese, D. L. Electrophysiological correlates of categorical speech perception in adults. *Brain and Language,* 1978, *5,* 25–35.

Molfese, D. L. The phoneme and the engram: Electrophysiological evidence for the acoustic invariant in stop consonants. *Brain and Language,* 1980, *9,* 372–376.

Molfese, D. L. Neural mechanisms underlying the processing of speech information in infants and adults: suggestions of differences in development and structure from electrophysiological research. In U. Kirk (Ed.) *Neuropsychology of language, reading, and spelling.* New York: Academic Press, in press.

Molfese, D. L., Erwin, R., & Deen, M. *Hemispheric discrimination of tone onset times by preschool children.* Paper presented at the annual meeting of the Psychonomic Society, St. Louis, Missouri, 1980.

Molfese, D. L., Freeman, R. B., Jr., & Palermo, D. S. The ontogeny of lateralization for speech and nonspeech stimuli. *Language and Brain,* 1975, *2,* 356–368.

Molfese, D. L., & Hess, R. M. Speech perception in nursery school age children: Sex and hemispheric differences. *Journal of Experimental Child Psychology,* 1978, *26,* 71–84.

Molfese, D., Laughlin, N., Morse, P., Linnville, S., & Wetzel, F. Neuroelectrical correlates of categorical perception for place of articulation in normal and lead-treated rhesus macaques. (Submitted)

Molfese, D. L., & Molfese, V. J. Hemisphere and stimulus differences as reflected in the cortical responses of newborn infants to speech stimuli. *Developmental Psychology,* 1979, *15,* 505–511. (a)

Molfese, D. L., & Molfese, V. J. Infant speech perception: Learned or innate. In H. A. Whitaker & H. Whitaker (Eds.), *Advances in neurolinguistics,* Vol. 4, New York: Academic Press, 1979. (b)

Molfese, D. L., & Molfese, V. J. Cortical responses of preterm infants to phonetic and nonphonetic speech stimuli. *Developmental Psychology,* 1980, *16,* 574–581.

Morse, P. Infant speech perception: A preliminary model and review of the literature. In R. Schiefelbusch & L. Lloyd (Eds.), *Language perspectives–acquisition, retardation, and intervention.* Baltimore: University Park Press, 1974.

Morse, P. The infancy of infant speech perception: The first decade of research. *Brain, Behavior, and Evolution,* 1979, *16,* 331–373.

Morse, P., Eilers, R., & Gavin, W. The perception of the sound of silence in early infancy. *Child Development,* 1982, *53,* 189–195.

Morse, P., Leavitt, L., Donovan, W., Kolton, S., Miller, C., & Judd-Engel, N. Headturning to speech: Explorations beyond the heart and the pacifier. *University of Wisconsin Infant Development Laboratory Research Status Report,* 1975, *1,* 347–351.

Morse, P., Molfese, D., Laughlin, N., Linnville, S. & Wetzel, F. Categorical perception for voicing contrasts in normal and lead-treated rhesus macaques: Electrophysiological indices. (Submitted)

Morse, P., & Snowdon, C. An investigation of categorical speech discrimination by rhesus monkeys. *Perception and Psychophysics,* 1975, *17,* 9–16.

Morse, P., & Suomi, K. Probing perceptual constancy for consonants with the non-nutritive sucking paradigm. *University of Wisconsin Infant Development Laboratory Research Status Report,* 1979, *3,* 311–319.

Sinnott, J., Beecher, M., Moody, D., & Stebbins, W. Speech sound discrimination by humans and monkeys. *Journal of the Acoustical Society of America,* 1976, *60,* 687–695.

Stevens, K., & Klatt, D. Role of formant transitions in the voiced-voiceless distinctions for stops. *Journal of the Acoustical Society of America,* 1974, *55,* 653–659.

Streeter, L. Language perception of 2-month-old infants shows effects of both innate mechanism and experience. *Nature,* 1976, *259,* 39–41.

Swoboda, P., Kass, J., Morse, P., & Leavitt, L. Memory factors in infant vowel discrimination of normal and at-risk infants. *Child Development,* 1978, *49,* 332–339.

Swoboda, P., Morse, P., & Leavitt, L. Continuous vowel discrimination in normal and at-risk infants. *Child Development,* 1976, *47,* 459–465.

Thomas, D., Campos, J., Shucard, D., Ramsay, D., & Shucard, J. Semantic comprehension in infancy: A signal detection analysis. *Child Development,* 1981, *53,* 798–803.

Till, J. *Infants' discrimination of speech and nonspeech stimuli.* Unpublished Ph.D. Dissertation, University of Iowa, Iowa City, Iowa, 1976.

Trehub, S. The discrimination of foreign speech contrasts by infants and adults. *Child Development,* 1976, *47,* 466–472.

Vargha-Khadem, F., & Corballis, M. *Cerebral asymmetry in infants.* Paper presented at the biennial meeting of the Society for Research in Child Development, New Orleans, Louisiana, 1977.

Warden, C., & Warner, L. The sensory capacities and intelligence of dogs, with a report on the ability of the noted dog ''Fellow'' to respond to verbal stimuli. *The Quarterly Review of Biology,* 1928, *3,* 1–28.

Werker, J., Gilbert, J., Humphrey, K., & Tees, R. Developmental aspects of cross-language speech perception. *Child Development,* 1981, *52,* 349–355.

COMMENTARY

What Sort of Psychophysics Is Infant Psychophysics?

Neil A. Macmillan

Department of Psychology
Brooklyn College
City University of New York
Brooklyn, New York

The experimental study of perception requires a psychophysics; that is, a set of procedures with which to pose questions and an array of models with which to formulate answers. The papers of Morse, Eilers and Oller, and Trehub, as well as those on basic auditory processes, reveal a community psychophysics for infant audition. In this essay, I hope to characterize this common approach and the manner in which it molds substantive conclusions. Although I will comment directly on each paper to some degree, most of what follows applies equally to all.

A concise answer to the title question is: Infant psychophysics is discrimination by untrained observers. Discrimination has been preferred to other techniques, I will argue, not only because it is simpler to apply, but for theoretical reasons which are most clearly set forth in Morse's paper. The psychophysical models adapted for use with infants were originally developed to describe trained and experienced observers; the second section of the paper considers the implications of applying such models to data from observers who are being studied precisely because they are inexperienced. Eilers and Oller are the most explicit about developmental effects, and their paper affords a context for discussing how the relation between psychophysical tasks may change. Finally, the domains of

Preparation of this chapter was supported by a grant from the PSC-CUNY Research Award Program of CUNY.

speech syllables and tone sequences are more complex than those studied by investigators of ''basic'' auditory processes. Trehub's contribution most clearly illustrates the particular difficulties and opportunities of doing psychophysics with structured stimuli.

Discrimination and Perception

The infant experiments in these chapters measure the ability to discriminate, using variants of the same–different design. Discrimination experiments have a correspondence function[1], which assigns to each possible stimulus an appropriate response. In most infant studies, the entire stimulus set during a condition consists of only two stimuli; the less common design in which the set of ''different'' stimuli (and sometimes the set of ''same'' stimuli) has more than one element is called an ''uncertain'' discrimination experiment. In adults, uncertain discrimination designs yield poorer performance than simple discrimination.

In a scaling task, the observers' responses provide information about the relative psychological magnitudes of stimuli or stimulus differences, and no correspondence function is usually defined. Of the many known scaling tasks (see Baird & Noma, 1978, for a primer), let me mention two. First, labeling, or identification tasks, require the subject to sort the stimulus set, which contains more than two elements, into at least two response categories. Second, difference judgment tasks require that responses be assigned to pairs of stimuli; the response is interpreted as a measure of similarity or difference. The best-known examples are the multidimensional scaling techniques developed originally by Shepard (see Shepard, Romney, & Nerlove, 1972).

Discrimination techniques have a number of practical advantages over scaling methods that have made them popular with students of infant and adult perception alike. The existence of a correspondence function allows feedback to be provided to the observer. Training effects in discrimination are apparently of shorter duration than in scaling, as I will argue in the next section. And models for extracting bias-free measures of sensitivity from discrimination data are well-established (Green & Swets, 1966). But discrimination data help little in answering such basic questions as, What is the infant's auditory world like? and, How does the infant's auditory system transform its input? To these questions, discrimination data can only say: (1) in the world of the infant, stimuli that cannot be discriminated are perceptually equivalent; and (2) if two stimuli are discriminable, they must at some point be processed differently. Scaling data would provide a more direct answer to at least the first of these questions.

[1]Bush, Luce, and Galanter (1963) used the term ''identification function'' for this relation, but this phrase is now commonly used to refer to performance in an identification experiment.

Of course, if discrimination and scaling were related in a predictable way, then studying discrimination would provide a short-cut to the scaling question. Although some psychophysicists have argued that the two tasks are essentially unrelated (Miller, 1956; Stevens, 1975), others have developed models according to which discrimination and scaling measure the same thing (Durlach & Braida, 1969; Thurstone, 1927). An especially influential version of this equivalence thesis is the hypothesis of "categorical perception," according to which many speech continua have the property that only stimuli that are identified as phonemically different can be discriminated (Liberman, Cooper, Shankweiler, & Studdert-Kennedy, 1967). Whether this actually occurs in adult speech perception experiments is doubtful, since attempts to predict discrimination from identification require additional assumptions (and free parameters) to succeed (Macmillan, Kaplan, & Creelman, 1977). Many investigators now describe as categorical any continuum in which peak discrimination occurs near the location of the phoneme boundary, as revealed by identification.

In his discussion of infant syllable discrimination, Morse refers to "categorical discrimination," a tacit acknowledgment that identification data do not exist for infants. To Morse, the discrimination data suggest nonetheless that speech sounds are processed by infants in a special mode, even in the first few postnatal months, and that this processing may well be innate, although not restricted to humans. The plausibility of this interpretation would be increased if we could assume that the infant's unmeasured identification performance is actually the same as that of an adult. Although this is a seductive assumption, I will argue in the next section that it should be viewed with suspicion. Eilers and Oller interpret the same data differently. According to their model, discrimination that is maximal at adult phoneme boundaries occurs even in the first stage of auditory development, a time at which only a basic auditory level of processing is available.

Both Morse and Eilers and Oller invoke the notion of "salience" in evaluating existing discrimination data, but offer different definitions of the term. For Morse, a contrast between speech sounds is salient if its discriminability is too great to be accounted for at a linguistic level; for Eilers and Oller, any contrast that is discriminable to Stage 1 infants is defined as salient. For stimuli that are not represented at multiple levels, salience is the same as discriminability; speech sounds have this characteristic during Stage 1, according to Eilers and Oller, and according to Morse, never. Salience is clearly a model-bound concept, but it will have to be bound more tightly if it is to be a testable one. Consider the problem of explaining the apparent decrease in sensitivity between infancy and adulthood to certain nonphonemic contrasts (e.g., Trehub, 1976). Morse suggests that salience changes with age, which reduces the usefulness of the concept. Eilers and Oller propose instead that access to Level 1 information that is not preserved at other levels becomes difficult; this idea retains the conceptual purity of salience, at the expense of an assumption that will be difficult to evaluate experimentally.

Whatever its ultimate definition, the idea of salience adds flexibility to the arsenal of speech perception theorists. The rise of this notion parallels the methodological shift in emphasis from heart-rate and sucking habituation designs to the head-turning paradigm. In typical experiments with habituation techniques, investigators ask whether the infant is able to discriminate between two syllables; the question is answered by seeking a statistically significant difference in response. Like all conventional statistical assessments, this decision procedure yields a binary outcome; binary data are congenial under the categorical-perception assumption, according to which discrimination will generally be all-or-none. The head-turning technique confronts the investigator with intermediate levels of discriminability, thereby encouraging the assessment of stimulus salience, presumably a continuous attribute.

Training and Development

As adult subjects perform in perception experiments, their performance improves; infant performance improves with development, as the few available cross-sectional studies show (e.g., Eilers, Wilson, & Moore, 1977) and common sense dictates. The most commonly drawn connection between these two effects concerns attempts to train adults to discriminate nonphonemic contrasts on speech continua (e.g., Carney, Widin, & Viemeister, 1977). I want briefly to describe another body of work on adult auditory learning, that of Watson and his colleagues (Watson, 1980; Watson & Kelly, 1981), which provides a useful model for interpreting developmental differences.

The stimuli in Watson's experiments are "word-length" tone sequences, and typically consist of ten 40-msec tone segments. They are thus similar to, but considerably more rapid than, the stimuli used by Trehub, and were chosen to approach the complexity of speech. Watson has measured simple discrimination, in which two specific sequences, often differing in only one element, are used, and uncertain discrimination, in which a change in one element is to be discriminated in the face of variability on irrelevant elements. Watson (1980) has also summarized a large number of studies, with different stimulus sets, on auditory identification. The most important findings, for purposes of the present analogy, are: (1) it takes longer to reach asymptotic performance in uncertain discrimination than in simple discrimination; and (2) it takes longer to reach asymptotic performance in identification than in discrimination. In Watson's most complex tasks, adult subjects continue to improve even after months of practice.

Watson and Kelly (1981) assume that simple discrimination reflects a "peripheral" process, whereas uncertain discrimination and identification draw on "central" factors. These terms are psychological, not physiological: uncertain discrimination and identification are central in that they require the application of cognitive skills, notably the allocation of attention. It is conventional in adult psychophysics to assume that uncertain discrimination is poorer than simple discrimination because of attention limitations (Swets, 1963), and that identification is poorer than discrimination because of either limited attention (Luce & Green, 1978) or increased memory load (Durlach & Braida, 1969).

The most obvious implication of this research for the development of speech perception is that infant identification performance cannot be inferred from discrimination. Identification improves slowly, relative to discrimination, so that even if the two tasks measure the same thing in adults (as the categorical perception hypothesis would have it), they cannot be equivalent in infants. Instead, discrimination gives an exaggerated impression of the overall perceptual ability of infants. Furthermore, longitudinal data on discrimination (particularly simple discrimination) after the neonatal period can be expected to yield artifically low estimates of the amount of learning taking place. The most significant perceptual ability acquired during development is not how to discriminate, but how to assign labels to (and otherwise scale) previously discriminable stimuli.

In general, this viewpoint fits well with the hierarchical scheme of Eilers and Oller, who would presumably expect identification to develop late because it uses levels of analysis not available earlier. An aspect of perceptual development addressed by Eilers and Oller but not captured in Watson's research is the finding that discrimination sometimes deteriorates with age. Eilers and Oller assume that, once Level 4 is reached, "the child may have difficulty directly accessing information at the phonetic level," where more acoustic information is retained than at the phonemic level. The fact that adults can learn to discriminate nonphonemic differences is attributed by Eilers and Oller to a late-developing ability to "gain conscious access to information" at low levels. This very interesting hypothesis has recently been put forward in more general terms by Gleitman (1979), who notes that the existence of multiple levels of analysis typifies many cognitive realms. Metacognitive skills, in this case the ability to report our speech percepts, are greatest, and develop earliest, for deep-level information. Titchener was aware of this: he found that subjects instructed to report surface aspects of perceptions or thoughts committed the "stimulus error," and described more meaningful aspects of their experience instead.[2]

The light shed by Watson's research on Trehub's investigations of tone-sequence discrimination is indirect. The long-term improvement of Watson's

[2]According to this argument, training strategies that are effective with adults will not necessarily work with infants. In the words of Chuck Watson (personal communication, 1982), infants are not just untrained listeners, they are not very bright untrained listeners. Adult auditory learning cannot be more than an imperfect model of infant auditory development, since both processes depend on the cognitive capacity of the listeners.

listeners, which is encouraging for the analogy between their perceptual improvement and that of infants learning to listen to speech, suggests that infant performance in Trehub's experiments may not improve much during the early years of life. Infants' (and most adults') exposure to tone sequences is slight compared to their exposure to speech, and there is little reason to expect them to develop cognitive tools just for perceiving them. The study of adult tone–sequence perception, notably that of Bregman (1981) on stream segregation, has been valuable in teasing apart some of the abilities required for speech perception, and interesting hypotheses about tone–sequence perception might arise from models of speech perception development. In terms of the Eilers and Oller model, for example, stream segregation and temporal grouping may be abilities that are required for syllabification to occur; if so, improvement in these tasks should occur at about the same time as in other Stage 2 tasks.

Scaling and Structure

Let me summarize the main argument. Scaling data provide the most direct answers to questions about perception, but infant psychophysics consists almost entirely of discrimination. This situation is generally thought to be acceptable because of the categorical-perception assumption that discrimination and identification measure the same thing. But this assumption cannot be correct for infants, even to the extent that it holds for adults, because the time course of perceptual learning in the two tasks is different. Thus, the picture of perceptual development provided by discrimination is distorted in at least two ways: (1) development seems to reach completion earlier than it in fact does; and (2) in infancy, discrimination is changing faster than perception generally; later, it is changing more slowly.

Clearly, it is important to develop scaling procedures that can be used with infants, and in this section I will describe two procedures that are little more than modifications of existing discrimination designs, but that at least provide a starting point. First, consider the identification experiment, which has actually been attempted for the special case (important in the speech–perception literature) in which there are only two responses (Aslin & Pisoni, 1980). A body of work that suggests the practicality of doing identification experiments is Kuhl's "constancy" research, described in all of the present papers. In a typical experiment, Kuhl (1976, 1980) trains infants to turn their heads in response to one phoneme but not another, in the face of irrelevant variation on another dimension (say, pitch or pitch contour). This procedure is most naturally viewed as uncertain discrimination, since each stimulus presentation has a corresponding correct response. But the same procedure could be used to study identification, using stimulus sets for which no correspondence function exists.

To my knowledge, no one has modified Kuhl's procedure for studying speech stimuli differing along a single physical dimension, although such an identification experiment would be entirely analogous to the two-response phoneme identification experiments commonly used with adults and animals (Kuhl & Miller, 1978).[3] The learning data summarized by Watson (1980) suggest that performance might be poor, but none of the experiments he describes was conducted with only two responses. In any case, it would seem straightforward and important to compare such infant identification experiments with the corresponding adult studies.

Let us return, however, to experiments with "multidimensional" stimulus sets, of which Kuhl's investigations are an important, but not a unique example. Multidimensionality also characterizes Trehub's tone sequences, which differ in key, temporal patterning, and melodic contour; and although many speech discrimination experiments are performed by attempting to select a set of stimuli that differ on one physical dimension, even single phonemes differ along several dimensions that have both articulatory and (in adults, at least) perceptual reality.

A scaling procedure that is often instructive when applied to such stimulus domains is multidimensional scaling (MDS). In MDS, measures of dissimilarity between each pair of stimuli are collected; the ordinal information in these measures can be used to locate the stimuli in metric spaces of varying dimensionality such that the interpoint distances come as close as possible to being monotonic with the input dissimilarities. (See Baird & Noma, 1978, for a detailed but accessible treatment.) The configuration whose dimensionality is high enough to yield low "stress" (violations of monotonicity) and low enough to have psychologically interpretable dimensions is treated as a representation of the listener's perceptual space for the stimulus set.

The input dissimilarities for MDS are often numerical judgments of difference, which are not available from infants, but confusion data can be used instead. Shepard (1972) analyzed confusion data collected by Miller and Nicely (1955) for 16 consonants; to insure that errors would be made, Miller and Nicely masked their stimuli with noise of different levels and bandwidth. In the resulting configuration, two dimensions, voicing and nasality, explained 99.4% of the variance. Cluster analysis, another multidimensional technique, revealed other aspects of this phoneme space.

Discrimination data obtained with infants could clearly be used as input to

[3]The two "responses" in infant studies are a head-turn and the absence of a head-turn; in adult and animal studies, the null response is not used. It has proved difficult to train infants to use two non-null responses (such as a left and right head-turn) in the same situation, but two-response identification could in principle use the same response set as discrimination.

MDS procedures. Proportion correct in a head-turning task may not be the ideal measure of discriminability, since it has a limited range; adaptive procedures, which have been used with infants (Aslin & Pisoni, 1980), could be applied to background noise level, or to some other correlate of task difficulty. In Trehub's experiments, discrimination is possible when sequences are presented right after one another; impossible if a sufficiently long interfering pattern is inserted between them. The duration of interference necessary for intermediate discriminability is a measure of similarity to which MDS could be applied.

Using discrimination data as the input to a scaling procedure subverts the discrimination/scaling distinction I have been drawing, and the MDS configuration derived from discrimination data may indeed be different from that derived from direct judgments or from identification data (which are what Miller and Nicely used). For adults, assuming the categorical perception hypothesis, the difference may be slight; for infants, in view of the task differences in the time course of auditory development, it could be substantial. Neither of the scaling procedures described here is ideal, but each can provide some information about the infant's auditory world that we do not now have.

Luce (1972) asked, "What sort of measurement is psychophysical measurement?" and answered that it was inherently inferior to physical measurement. Infant psychophysics is in a much better situation: that it cannot match adult psychophysics in every detail is most naturally attributed to its youth. Infant psychophysicists have been very creative in devising questions to ask their listeners, and, as these chapters show, increasingly sophisticated in elaborating interpretive models. A necessary condition for continued growth is that methodology and theory mature together.

References

Aslin, R. N., & Pisoni, D. B. Some developmental processes in speech perception. In G. H. Yeni-Komshian, J. F. Kavanaugh, & C. A. Ferguson (Eds.), *Child phonology. Vol. 2. Perception.* New York: Academic Press, 1980.

Baird, J. C., & Noma, E. *Fundamentals of scaling and psychophysics.* New York: Wiley, 1978.

Bregman, A. S. Asking the "What for" question in auditory perception. In M. Kubovy & J. R. Pomerantz (Eds.), *Perceptual organization.* Hillsdale, N.J.: Erlbaum, 1981.

Bush, R. R., Luce, R. D., & Galanter, E. Characterization and classification of choice experiments. In R. D. Luce, R. R. Bush, & E. Galanter (Eds.), *Handbook of mathematical psychology Vol. 2.* New York: Wiley, 1963.

Carney, A. E., Widin, G. P., & Viemeister, N. F. Noncategorical perception of stop consonants differing in VOT. *Journal of the Acoustical Society of America,* 1977, *62,* 961–970.

Durlach, N. I., & Braida, L. D. Intensity perception. I. Preliminary theory of intensity resolution. *Journal of the Acoustical Society of America,* 1969, *46,* 372–383.

Eilers, R. E., Wilson, W. R., & Moore, J. M. Developmental changes in speech discrimination in infants. *Journal of Speech and Hearing Research,* 1977, *20,* 766–780.

Gleitman, H. *Some trends in the study of cognition.* Paper presented at the convention of the American Psychological Association, New York, 1979.

Green, D. M., & Swets, J. A. *Signal detection theory and psychophysics.* New York: Wiley, 1966.

Kuhl, P. K. Speech perception in early infancy: The acquisition of speech-sound categories. In S. K. Hirsh, D. H. Elderdge, I. J. Hirsh, & S. R. Silverman (Eds.), *Hearing and Davis.* St. Louis: Washington University Press, 1976.

Kuhl, P. K. Perceptual constancy for speech-sound categories in early infancy. In G. H. Yeni-Komshian, J. F. Kavanaugh, & C. A. Ferguson (Eds.), *Child phonology. Vol. 2. Perception.* New York: Academic Press, 1980.

Kuhl, P. K., & Miller, J. D. Speech perception by the chinchilla: Identification functions for synthetic VOT stimuli. *Journal of the Acoustical Society of America,* 1978, *63,* 905–917.

Liberman, A. M., Cooper, F. S., Shankweiler, D. P., & Studdert-Kennedy, M. Perception of the speech code. *Psychological Review,* 1967, *74,* 431–461.

Luce, R. D. What sort of measurement is psychological measurement? *American Psychologist,* 1972, *27,* 96–106.

Luce, R. D., & Green, D. M. Two tests of a neural attention hypothesis for auditory psychophysics. *Perception and Psychophysics,* 1978, *23,* 363–371.

Macmillan, N. A., Kaplan, H. L., & Creelman, C. D. The psychophysics of categorical perception. *Psychological Review,* 1977, *84,* 452–471.

Miller, G. A. The magical number seven, plus or minus two: Some limits on our capacity for processing information. *Psychological Review,* 1956, *63,* 81–97.

Miller, G. A., & Nicely, P. E. An analysis of perceptual confusions among some English consonants. *Journal of the Acoustical Society of America,* 1955, *27,* 338–352.

Shepard, R. N. Psychological representation of speech sounds. In E. E. David & P. B. Denes (Eds.), *Human communication: A unified view.* New York: McGraw-Hill, 1972.

Shepard, R. N., Romney, A. K., & Nerlove, S. B. *Multidimensional scaling. Vol. 1. Theory.* New York: Seminar Press, 1972.

Stevens, S. S. *Psychophysics.* New York: Wiley, 1975.

Swets, J. A. Central factors in auditory frequency selectivity. *Psychological Bulletin,* 1963, *60,* 429–440.

Thurstone, L. L. A law of comparative judgment. *Psychological Review,* 1927, *34,* 273–286.

Trehub, S. E. The discrimination of foreign speech contrasts by infants and adults. *Child Development,* 1976, *47,* 466–472.

Watson, C. S. Time course of auditory perceptual learning. *Annals of Otology, Rhinology, and Laryngology,* 1980, *89,* suppl. 74, 96–102.

Watson, C. S., & Kelly, W. J. The role of stimulus uncertainty in the discrimination of auditory patterns. In D. J. Getty & J. H. Howard (Eds.), *Auditory and visual pattern recognition.* Hillsdale, N.J.: Erlbaum, 1981.

Index